BASIC
ELECTROTECHNOLOGY

H. COTTON, D.Sc.
Emeritus Professor of Electrical Engineering
University of Nottingham

MACMILLAN

© H. Cotton 1973

First published 1973 by
THE MACMILLAN PRESS LTD
London and Basingstoke
Associated companies in New York
Dublin Melbourne Johannesburg and Madras

SBN 333 14311 6

Printed in Great Britain by
MORRISON AND GIBB LTD
London and Edinburgh

PREFACE

Unlike a treatise, which is a statement of original work undertaken to increase knowledge, whether that knowledge can be applied to human problems or not, a text-book is a collection of material already known.

The justification for an additional text-book to the vast number already in existence is therefore dependent on (a) the choice of material and its suitability to some specific purpose, and (b) the clarity of the exposition; is it such that the book can be understood by students of average ability?

The present book satisfies requirement (a) because it is written for a specific purpose, namely, to cover the syllabus of the subject of Electrical Principles II of the City and Guilds of London Institute. As far as requirement (b) is concerned it is hoped that the author's long experience of technical writing has ensured that this also is satisfied.

Where it seemed advantageous the requirements of the syllabus have been exceeded in a few cases, for example, a discussion of the present-day importance of the metal sodium as a conductor of electricity, and a description of the mercury cell.

The book ends with a collection of exercises, and since it is probable that the so-called Imperial system of units will survive for a few years the data for a few of the exercises corresponds to this system.

As this is a new book there may be suggestions, or even criticisms, from readers. The author will be grateful for such co-operation.

Woodbridge,
Suffolk

H. COTTON

CONTENTS

CONTENTS

1 THE SI SYSTEM OF UNITS

The present advanced state of the pure and applied sciences could not have been achieved without experimentation and without measurements made with the utmost accuracy possible with existing resources. By measurement we mean comparison with something of the same physical nature which is regarded as a standard. Hence, in order to state the magnitude of any physical entity we make use of two concepts, a numeric and a unit. A numeric is a number and it may be integral or fractional, positive or negative. A unit is a physical entity of such magnitude that it has a numeric of unity. Thus if we say that an electric current is equal to 10 amperes, 10 is the numeric and ampere is the name given to a current of such magnitude that it is equal to the accepted unit of current.

We are concerned with two kinds of unit; the fundamental or base unit, and the derived unit. A base unit is one which can only be defined in terms of itself; for example a length can only be stated in terms of another length. Hence the unit of length is a base unit. On the other hand the unit of velocity, which is given by the rate of movement, is a derived unit because velocity involves both distance and time.

In the SI system (Système International d'Unités) the base units are as follows:

Length The unit is the metre (m)
Mass   The unit is the kilogram (kg)

It is very important at the outset to distinguish between mass and weight. Mass is the amount of matter in a body and weight is the force with which that particular mass is attracted by the gravitational system in which it happens to be placed. For example, if a piece of material of mass 1 kg is taken from the earth to the moon, its mass will remain at 1 kg but its weight will have a number of values. On the earth it will be one kg weight; during orbit with the engines not acting, that is during free flight, its weight will for most practical purposes be zero; on the moon its weight will be one-sixth of a kg.

Time   The unit of time is the second (s). It is equal to one 86 400th part of the mean solar day.

The above base units are not sufficient to define electrical entities and it has proved necessary to introduce a base unit of purely electrical nature. This is the unit of electric current, the ampere (A). From the following definition it will be seen that its function is to provide a bridge between electrical and mechanical entities.

Electric current

The ampere is that current which, when flowing in each of two infinitely long parallel conductors *in vacuo*, and separated one metre between centres, causes each conductor to experience a mechanical force of 2×10^7 newton per metre length. (The newton is the SI unit of force and is defined below.)

Temperature

The unit is the kelvin (K), previously called the degree absolute. It is numerically equal to the familiar degree centigrade but the cardinal points are different. Thus, 0° centigrade, 0° C, is equal to 273 K and 100° C is equal to 373 K. The absolute zero of temperature, 0 K, is thus $-273°$ C.

Luminous Intensity

The unit is the *candela* (cd). We are not concerned with this unit in this book.

Derived Units

Derived units of electrical nature will be defined in appropriate places in the text; the following units in the mechanical system will be used:

Velocity is defined as the ratio of a linear distance travelled to the time taken to cover that distance. Thus if the velocity v is uniform during time t and the distance covered is d then

$$v = d/t$$

and the SI unit is the metre per second, m/s.

Acceleration is the rate of increase of velocity. An increase in velocity is of the same nature as a velocity, just as an increase in length is of the same nature as a length.

$$a = (\text{change in velocity})/t$$

and the SI unit is metre per second per second, m/s².

Example 1.1 A body travelling at 10 m/s accelerates uniformly and reaches a speed of 30 m/s in one hour. What is its acceleration?

$$\text{Change in velocity} = 30 - 10 = 20 \text{ m/s}$$
$$\text{Time} \quad t = 60 \times 60 = 3600 \text{ s}$$
$$\therefore \ a = 20/3600$$
$$= 0.0056 \text{ m/s}^2$$

If u and v are the initial and final velocities respectively, a the acceleration, d the distance travelled and t the time, the equations of motion for a body or particle moving with uniform acceleration are:

$$v = u + at$$
$$v^2 = u^2 + 2\,ad$$
$$d = ut + \tfrac{1}{2}\,at^2$$

Force

If a mass m is subjected to an acceleration a through application of a force, then

$$F = ma$$

If $m = 1$ kg and $a = 1$ m/s² then $F = 1$ SI unit, and the name of this unit is the newton (N). Hence the newton is that force which, acting on a mass of 1 kg produces an acceleration of 1 m/s².

Example 1.2. When an electron of mass 9.1×10^{-28} g is placed in a certain electric field the force acting on it is 2.5×10^{-15} N. Calculate its acceleration.

$$a = F/m$$
$$F = 2.5 \times 10^{-15} \text{ N}$$
$$M = 9.1 \times 10^{-28} \text{ g} = 9.1 \times 10^{-31} \text{ kg}$$
$$\therefore \quad a = \frac{2.5 \times 10^{-15}}{9.1 \times 10^{-31}}$$
$$= 2.747 \times 10^{15} \text{ m/s}^2$$

If a body is falling freely under the action of gravity, the acceleration due to the force of gravity is equal to 9.81 m/s and acts directly downwards. The force of gravity acting on a mass of m kg is therefore

$$F = m\,a = 9.81\,m \text{ newtons}$$

Work and Energy

These are names for the same thing; we say that energy is expended when work is done. If a force F acts over a distance d then the work done is

$$W = F\,d$$

and the SI unit is the newton metre (N m). We shall see that this is the same as a unit called the joule (J).

Example 1.3.   Calculate the work done, and therefore the energy acquired by the electron in example 2 after it has moved a distance of 2 cm under the influence of the electric field.

$$d = 2 \text{ cm} = 2 \times 10^{-2} \text{ m}$$
$$F = 2.5 \times 10^{-15} \text{ N}$$
$$\therefore \quad W = F\,d$$
$$= 2.5 \times 10^{-15} \times 20 \times 10^{-3} = 50 \times 10^{-18} \text{ N m or J}$$

Angular Velocity

Electrical machines are rotating machines and therefore the concept of angular velocity is important. Linear velocity is equal to linear distance divided by time; similarly angular velocity is equal to angular distance divided by time. Suppose that a particle is travelling in a circular path. From figure 1.1, let it move from A to B, in time t, and let the angle swept out as the radius OA moves to OB be θ; then

$$\text{Angular velocity} = \theta/\text{time}$$

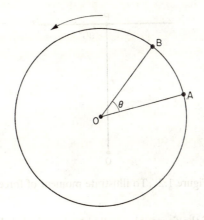

Figure 1.1 To illustrate angular velocity.

The unit of angle is the radian (rad), and the symbol for angular velocity is ω

$$\therefore \; \omega = \theta/t$$

and the SI unit is the rad/s.

When the point has made one revolution the radius has swept out 2π radians. If the time of one revolution is T, then

$$\omega = 2\pi/T$$

Example 1.4.   A turbo-alternator rotates at 3000 rev/min. What is its angular velocity?

$$\begin{aligned}
\text{Speed} \quad n &= 3000 \text{ rev/min} \\
&= 50 \text{ rev/s} \\
\therefore \; T &= 1/n = 1/50 = 0.02 \text{ s} \\
\therefore \; \omega &= 2\pi/0.02 = 100\pi \\
&= 314 \text{ rad/s}
\end{aligned}$$

Moment of Force

Figure 1.2 shows a line OP of length r which can rotate about its end O. A force F acts at P, its direction being at right-angles to OP.

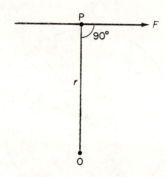

Figure 1.2   To illustrate moment of force.

The moment of the force, also called torque, is equal to the product Fr. The SI unit is therefore the newton metre (N m).† If we denote the moment of the force by T, then we have the equation

$$T = Fr$$

Note that we have used the same symbol T for the entity time as for the entity moment of force. As there are so many entities, duplication of symbols can hardly be avoided, but the nature of the context does away with possible confusion.

Power is the rate of doing work. Thus if W newton metres of work are performed in time t, the power is given by

$$P = \text{work/time}$$
$$= W/t \text{ N m/s}$$

† It will be noticed that the unit of torque and the unit of work or energy are the same, namely the newton metre, N m. This is because both are given by the product of a force and a length. Physically they are quite different because motion is necessary in order that work may be done by the direct application of a force, but a torque can be exerted even if there is no movement. For this reason it is preferable to use N m to denote the magnitude of a torque and J to denote the work done, or energy.

Thus the SI unit is the newton metre per second; we shall see that it is the same as the unit called the watt.

Example 1.5. A pump raises 454 kg of water through a height of 30 m in 5 minutes. Calculate the work done and the power.

$$m = 454 \text{ kg}$$
$$\therefore \text{ Force of gravity } F = mg$$
$$= 454 \times 9.81$$
$$= 4454 \text{ N}$$
$$d = 30 \text{ m}$$
$$\therefore W = Fd$$
$$= 4454 \times 30$$
$$= 1.336 \times 10^5 \text{ N m or J}$$
$$\text{Time taken } t = 5 \text{ min} = 300 \text{ s}$$
$$\therefore P = \frac{1.336 \times 10^5}{3 \times 10^2}$$
$$= 445.3 \text{ N m/s or W}$$

Consider again figure 1.2. Suppose that, under the influence of the force F, which always maintains its direction at right-angles to PO, the line PO is made to rotate about O. In one revolution the linear distance moved is equal to the circumference

$$d = 2\pi r$$
$$\therefore \text{ Work done } W = 2\pi (r \times F) = 2\pi T$$

If the time taken is t, then

$$\text{Power } P = W/t = (2\pi/t) T$$

But $2\pi/t$ is the angular velocity ω

$$\therefore P = \omega T$$

Example 1.6. The shaft output of an electric motor is 60 000 N m/s and its speed is 720 rev/min. Calculate the shaft torque.

$$\text{Speed } n = 720/60 = 12 \text{ rev/s}$$
$$\therefore \omega = 12 \times 2\pi = 75.5 \text{ rad/s}$$
$$\therefore T = \frac{P}{\omega} = \frac{60\,000}{75.5}$$
$$= 794.7 \text{ N m}$$

The Equations of Motion

The science of electronics is largely concerned with the motions of particles, the forces acting on them, their accelerations and their velocities. There are three fundamental equations and it is preferable to derive them in terms of the motions of more familiar bodies.

Let m = mass of body
F = force acting on it
a = acceleration due to the force f
u = initial velocity
v = velocity after time t

The above entities are all expressed in SI units.

1. Let a body experience a uniform acceleration a. After a time t has elapsed the velocity will have increased above the initial velocity by the amount at, so that, with a body starting from rest ($u=0$) the velocity after time t will be at. If the force which produces the acceleration is applied to a body which already has a velocity u, then after time t, the velocity will be

$$v = u + at \qquad (1)$$

Example 1.7. A body has an initial velocity of 10 m/s. If it is falling freely under gravity, its attraction then being $g=9.81$ m/s², what will be its velocity after 5 seconds?

$$a = g = 9.81 \text{ m/s}^2$$
$$\therefore \quad v = 10 + 9.81 \times 5$$
$$= 59.05 \text{ m/s}$$

2. Velocity in terms of acceleration and distance moved. First let $u=0$ and let a be constant. After an elapsed time t the velocity will be $v=at$. But since the initial velocity is zero and the acceleration is uniform the *average* velocity is $\frac{1}{2}v=\frac{1}{2}at$ and the distance moved

$$d = \text{average velocity} \times \text{time}$$
$$= \frac{1}{2}at \times t = \frac{1}{2}at^2$$

If the body had an initial velocity u and no acceleration the distance moved would be ut. For a body with a finite velocity as well as acceleration we therefore have

$$d = ut + \frac{1}{2}at^2 \qquad (2)$$

Example 1.8. How far will the body of example 1.7 have fallen during the 5 seconds?

$$a = g = 9.81 \text{ m/s}$$
$$\therefore \quad d = 10 \times 5 + \tfrac{1}{2} \times 9.81 \times 5^2$$
$$= 172.6 \text{ m}$$

3. Final velocity in terms of initial velocity, acceleration and distance travelled.

$$v = u + at$$
$$\therefore \quad v^2 = u^2 + (at)^2 + 2uat$$
$$v^2 = u^2 + 2a(ut + \tfrac{1}{2}at^2)$$
$$v^2 = u^2 + 2ad \qquad (3)$$

Example 1.9. A body is projected vertically upwards. What is its initial velocity in order that it may reach 50 m? If it then falls to the ground how long does it take?

During ascent the acceleration must be taken as negative because it produces a progressive diminution in velocity.

$$\therefore \quad v^2 = u^2 - 2gd$$
$$0 = u^2 - 2 \times 9.81 \times 50$$
$$u^2 = 981$$
$$u = 31.3 \text{ m/s}$$

During descent the acceleration is again positive and we now take u as zero and the final velocity as 31.3 m/s.

$$\therefore \quad v = u + gt$$
$$31.3 = 0 + 9.81t$$
$$t = 31.3/9.81 = 3.2 \text{ s}$$

Application of the above equations to the motion of charged particles placed in electric fields are given in chapter 18.

Kinetic Energy

Suppose that a mass m is raised from some datum level through a vertical distance d to a second, higher, level. Then the work done against the attraction of gravity will be mgd, the SI unit being the joule. This amount of work will be done, whatever the path taken between the two levels. The criterion is the vertical distance d between them. Thus work, and therefore energy, are not directional and are therefore not vector quantities; they are scalar quantities characterised by a numeric and the name of the unit.

The whole of the work done will now be in the form of the potential energy of the mass by virtue of its new position.

Now suppose that the mass is released and can fall freely under gravity. It will be subject to the acceleration g, so that its velocity will increase. This velocity will impart energy of motion, kinetic energy, and at any instant the kinetic energy will be equal to the loss of potential energy due to the fall to the position at that particular instant. Hence after a fall of d, the kinetic energy will be equal to mgd. If the mass is released with an initial velocity of zero, the distance d in terms of the velocity acquired is

$$v^2 = 2\,gd$$
$$d = v^2/2\,g$$

Hence, denoting the kinetic energy by W

$$W = mg \times (v^2/2\,g)$$
$$W = \tfrac{1}{2}\,mv^2$$

This is independent of the manner in which the velocity v is acquired and is independent of the direction of v since W is a scalar quantity. Thus, although for simplicity, we assumed an initial velocity of zero, the expression is perfectly general, the kinetic energy for a given mass at any instant depending only on the velocity at that instant.

Example 1.10.   A mass of 2 kg, starting from rest, falls freely through a vertical distance of 100 m. What velocity does it acquire, and what is its kinetic energy?

$$v^2 = 0^2 + 2 \times 9.81 \times 100$$
$$= 1962$$
$$v = 44.3 \text{ m s}^{-1}$$
$$W = \tfrac{1}{2} \times 2 \times (44.3)^2$$
$$= 1962 \text{ J}$$

Energy

Einstein showed that energy and matter are interchangeable in the sense that one can be converted into the other. Thus matter is converted into energy during atomic and nuclear reactions, and there is a definite relationship between the mass of the matter so used and the energy in joules into which it is converted. We can thus regard energy as a constituent of the universe, a constituent of paramount importance to the engineer.

According to the law of conservation of energy, the creation or destruction of energy is not possible. What is possible is the conversion of one form of energy to another. Since matter can be converted into energy we have to include mass in the statement of the law of conservation.

The electrical engineer is concerned with energy in the forms detailed below.

1. Potential energy due to position within a field of force. The most familiar example of a field of force is the gravitational field of the earth. If a mass m rests on the surface of the earth we say that its potential is zero *with respect to the surface*. If it is raised through a vertical distance h then work equal to mgh has to be performed against the force of gravity, and this amount of work is the potential of the body in its new position. We shall see that an electric charge has a potential in virtue of its position in an electric field of force.

2. Kinetic energy in virtue of velocity. We have seen that the kinetic energy of a mass is equal to $\frac{1}{2}mv^2$. A body at rest therefore has zero kinetic energy.

3. The energy stored in a field of force. We shall see that energy is stored in both electric and magnetic fields of force. As a simple example imagine the north and south poles of two bar magnets to be in contact. If they are separated by the application of a force which overcomes their mutual attractions, work will be done equal to the product of the applied force and the distance of separation measured in the direction of that force. At the same time a magnetic field of force will be set up between the poles and energy will be stored in this field equal to the amount of work done in effecting the separation.

Summary

We have used two sets of symbols, those for the entities and those for the si units. The symbols for the units must not be changed but it is sometimes necessary to use several symbols for the same entity. Thus with length, the si unit has the symbol m but a length (not the unit), may be denoted by l, d or s. Similarly with other entities.

Entity	Symbol for entity	Name of unit	Symbol for unit
		Base units	
Length	l, d, s	metre	m
Mass	m	kilogram	kg
Time	T, t	second	s
Electric current	I, i	ampere	A
Temperature	T, θ	Kelvin	K
		Derived units	
Velocity	u, v	metre per second	m/s
Acceleration	a	metre per second per second	m/s^2
Force	F	newton	N
Work, energy	W	metre newton or joule	m N or J
Power	P	metre newton per second, or watt	m N/s or W
Angular velocity	W	radian per second	rad/s
Moment of force (or torque)	T	newton metre	N m

2 ELECTRICITY

Electricity is a constituent of all matter in its normal state. There are two kinds of electricity called positive and negative and if they are brought together they tend to neutralise one another's effects in space between them. If two quantities of electricity, one positive the other negative, completely neutralise one another's effects then the amounts of each are said to be equal. Since matter in its normal state does not exhibit any state of electrification it follows that the amounts of the two kinds of electricity in it are equal.

The smallest amount of an element which retains the properties of that element is the atom. The two kinds of electricity are parts of the normal atom so that the atom contains equal amounts of the two kinds as well as matter in the macroscopic sense, that is, matter in the bulk. With one exception the atoms also contain non-electric matter. Thus, in the majority of atoms there are the three substances and it is usual to regard them as particles because they behave like particles. They are

Protons: particles of positive electricity
Electrons: particles of negative electricity
Neutrons: neutral particles

The protons and neutrons can be regarded as permanent constituents of the atom since they cannot be removed by the normal processes of electrical engineering. The electrons, however, can be displaced from their normal positions relative to the protons and neutrons, and, under appropriate conditions, one or more can be removed altogether. If this happens, then the electrical balance of the atom is destroyed and it becomes electrically positive instead of neutral. An atom which loses one or more electrons is called a positive ion. Figure 2.1 shows diagrammatically the arrangements of

13

the electrons and protons in the normal atoms of nine elements. The first is the hydrogen atom which has one proton but no neutron for its nucleus, and one electron. The electrons are associated with the nucleus in a very definite manner, namely, in groups which are

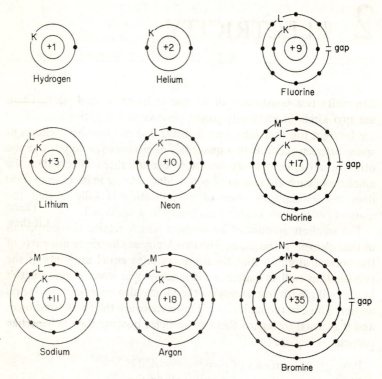

Figure 2.1   Some atomic electron arrays.

The first column shows hydrogen and two of the alkali metals. Hydrogen, with its single electron vehaves, chemically, like a metal.

The second column shows three of the inert gases, their chemical inertness resulting from complete electron shells.

The third column shows three of the halogen gases. These are characterised by a gap of one electron in the outermost shell. Because of this they are chemically very active. It will be seen that the M shells for argon and bromine are shown with different numbers of electrons for completion. This is because some of the shells consist of two, or three, subshells. The M shell consists of three such subshells, their electron numbers being 2, 6 and 10. Thus argon has $2+6=8$ electrons in the M shell and bromine has $2+6+10$, that is 18 in all.

called shells. We see that the normal hydrogen atom has one electron in what is called the 'K' shell.

The second example is the helium atom, which has two protons and two neutrons in its nucleus, and two electrons in the K shell. In all normal atoms there are as many extra nuclear electrons as there are nuclear protons. The K shell can only contain two electrons; so the next element, lithium, which has three electrons must have one of these in the next shell, the L shell. There are also three protons in the nucleus and the number of neutrons is increased to four. The neutrons are not shown.

Proceeding in this way we can build up models of all the elements; thus the seventh diagram shows the electron array of the sodium atom. It has two electrons in the K shell, eight in the L shell, and one in the next, the M shell. Thus it has eleven electrons altogether and therefore there are eleven protons in the nucleus. We need no longer concern ourselves with the neutrons as they do not take part in any of the phenomena described in this book.

Ionisation

As we have seen, an atom becomes a positive ion if it loses one or more of its electrons. We shall denote the quantity of electricity in the electron and the proton by e. If a hydrogen atom loses its single electron it becomes a hydrogen ion with a charge of $+e$ (figure 2.1). If the sodium atom loses its single electron in the M shell it, also, becomes an ion with a charge of $+e$. It is only the electrons in the outermost shell which can be shed in this way, in fact it is the composition of this shell which largely decides the physical and chemical properties of the element.

In general the metallic elements have only one or two electrons in the outermost shell; for example copper and silver have one, zinc and mercury have two. Aluminium is an exception, having three. It is this small number of outermost electrons which give these metals their good electrical conducting properties. The hydrogen atom, having only one electron, is like a metal atom, its chemical behaviour being very similar to metals such as sodium, copper and silver.

It is useful to have some idea of the magnitude of the electron: it behaves as though it were a particle of mass 9.107×10^{-31} kg, and of radius 10^{-15} m.

Electrification

Consider a rod of a monovalent element such as copper; by mono-valent we mean having only one outermost electron to each atom. These outermost electrons are so weakly held by the attraction of the nucleus that they are able to escape and wander about in the relatively vast spaces between the atoms, which have now become positive ions (figure 2.2). Electron motion is largely decided by the temperature and it is therefore called random thermal motion. The ions cannot wander about in this way although, as we shall see, they are able to oscillate about a fixed mean position. Metals such as copper are good conductors of electricity for two main reasons: first, the outermost, or valence, electrons escape easily from the atom and can wander about freely; the number of electrons is extremely large.

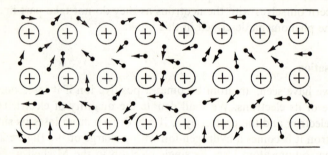

Figure 2.2 The ions of a metallic conductor with the electron gas wandering in the interspace.

Thus in copper there are 2×10^{28} free electrons per m³; second, like electricity, such as two negatives or two positives, repel one another; unlike electricities, that is positive and negative, attract one another. Insulators are extremely bad conductors of electricity, because they have very few free electrons. They can, however, be electrified. Their properties are discussed in Chapter 14.

Electrification by Friction

If two different bodies are rubbed violently together they will both become electrified because electrons from one will be transferred to the other, one thereby becoming positively and the other negatively electrified. This is not merely a scientific curiosity but is of fairly common occurrence. Thus in desert regions the friction of sand on telephone lines can raise the wires to a very high voltage; 11 000V has been recorded. Escaping high pressure steam from leaking glands of steam engines can have a similar effect; in fact, one of the earliest electrical machines was a boiler mounted on insulating supports. Steam escaped through a row of small nozzles and the whole boiler thereby became electrified.

Electrification by Induction

Figure 2.3 shows a conducting bar which is insulated from earth although, for simplicity, the insulating support is not shown. A body having positive electrification, usually called a positive charge,

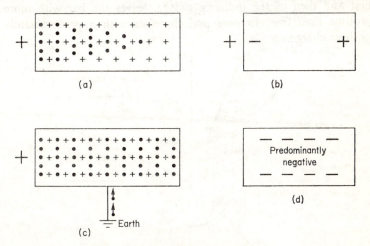

Figure 2.3   Electrification by induction.

is placed near the left-hand end. It repels the protons in the bar and attracts the electrons. But only the free electrons can move and, as a result, electrons are drained from the right-hand and accumulate at the left-hand end (figure 2.3a). The net result is predominantly negative electrification at the left-hand and predominantly positive electrification at the right-hand, as indicated very simply in figure 2.3b. So far there has been no change in the total number of electrons in the bar.

Figure 2.3c shows the effect of connecting the bar to earth by means of a conducting wire. The earth is so vast that it can supply or receive electrons without any appreciable change in itself. In a sense the bar and the earth are now one conducting system and the positive inducting charge causes electrons to pass from earth to the bar. If the inducing charge is now removed, as in figure 2.3d, the additional electrons received from the earth distribute themselves uniformly within the bar which, in consequence, becomes negatively charged.

If the inducing charge is negative then the condition corresponding to figure 2.3a will be one of uniform distribution of positive charges but with the electrons repelled towards the right-hand end. If an earth connection is now made, the inducing charge remaining in place, and electrons from the bar will escape to the earth, resulting in a deficit of electrons in the bar. The removal of the earth connection first and then of the inducing charge, leaves the bar with more protons than free electrons and therefore with a predominantly positive charge.

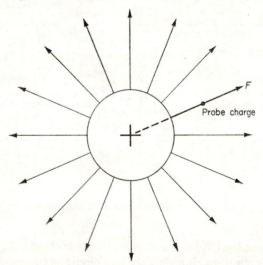

Figure 2.4   Electric field due to an isolated positively charged sphere.

When an object is 'charged' we mean that it has received something. The term charge is therefore appropriate when the electrification is negative but not when it is positive, because positive electrification involves the removal, not the addition, of electrons.

The Electric Field. Lines of Force

Imagine an insulated body, a sphere for simplicity, having a positive charge. Imagine also a minute positive charge placed near to it, and let this charge be so extremely small that it has no appreciable effect on the space surrounding the sphere as in figure 2.4. Such a minute charge is called a probe charge. The probe charge will be repelled, and clearly the force F acting on it will be directed along a radius of the sphere. This will apply whatever the position of the probe charge and whatever its distance from the sphere. One extended radius of the sphere is a line for which at every point on it the direction of the line is the direction of the force on the probe charge. Such a line is called a line of force. The whole of the region throughout which there is a detectable force on the probe charge is called the *electric field* of the spherical charge. Clearly the electric field of an isolated spherical conductor consists of radial straight lines. With a positive charge the force on the probe charge is outwards and therefore the lines of force can be arrowed outwards as shown.

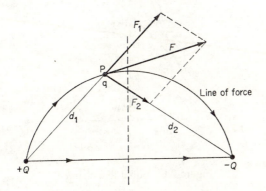

Figure 2.5   Equal unlike charges.

Consider two equal neighbouring charges, one positive, one negative. Because of radial symmetry we see that, external to itself a charged sphere acts as though its charge is concentrated at its centre. Consequently the shape of the field is the same as that due to a point charge placed at the centre. This also applies if the charged bodies are not spherical, except for points close to the bodies. We will therefore consider two opposite point charges, figure 2.5. Let each be of magnitude Q, leaving for a while any consideration of the unit of quantity of electricity, and let the magnitude of the probe charge be q. If q is distant d from Q then the force on q is given by

$$F = \frac{1}{k} \frac{Qq}{d^2}$$

where k is a constant whose value depends on the medium surrounding the charges. This is in accordance with the inverse square law which states that the force of attraction, or repulsion between two small charges is proportional to their product and inversely proportioned to the square of the distance between them.

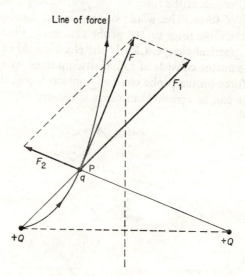

Figure 2.6   Equal like charges.

In order to determine the 'shape' of the field we are concerned only with relative values so that

$$F \propto \frac{Qq}{d^2}$$

With Q and q given and of constant magnitudes

$$F \propto \frac{1}{d^2}$$

Let the probe be at any point P; then it is repelled by $+Q$ by a force F_1 acting along Q⁺P produced, and it is attracted by $-Q$ by a force F_2 acting along PQ⁻.

$$\therefore \frac{F_1}{F_2} = \frac{(d_2)^2}{(d_1)^2}$$

The total force F acting on the probe is the resultant of F_1 and F_2 as given by the diagonal of the parallelogram of forces. The direction of F is tangential to the line of force which passes through P and this gives the direction of the line of force at that point. By taking a series of positions for P the directions of the lines of force through those points can be determined and the map of the field then sketched in.

In the case of two equal neighbouring charges, both positive, the probe charge is now repelled by both charges, so that the construction becomes that of figure 2.6. With unlike charges the lines of force start from the positive charge and end on the negative. With like positive charges they start from both charges and end, theoretically at infinity, but actually at conducting objects in their vicinity or at the earth's surface. With like negative charges they all end at these charges.

Figure 2.7 shows the shapes of the lines of force for four fields. Case 4 is that of a long conductor parallel to the earth's surface and insulated from it. The lines of force terminate normally to the earth's surface and we can therefore imagine that they are continued on the other side of this surface as shown dotted, somewhat after the manner of an optical image in a mirror. The field between conductor and earth is identical with that between two equal and unlike charged parallel wires whose distance apart is equal to twice the height of the wire.

There can be no electric lines of force inside the earth itself. We shall see in Chapter 3 that unless an electric current is flowing there can be no electric field inside a conductor. If a conductor is hollow and charged bodies insulated from it are introduced, then there will be a field due to these bodies.

Summary of the properties of Electric Lines of Force

1. An electric line of force begins at a positive charge and ends at a negative charge.
2. There are no lines of force inside a hollow charged conductor unless there are insulated charged bodies inside it.

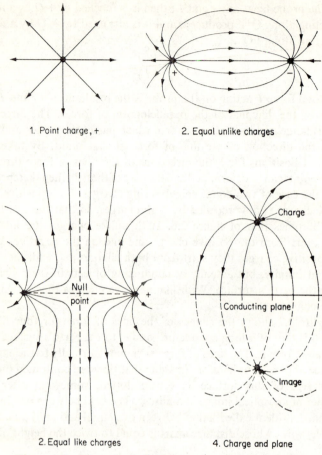

1. Point charge, +   2. Equal unlike charges

2. Equal like charges   4. Charge and plane

Figure 2.7   Electrostatic field forms in a few typical cases.

3. Two lines of force cannot intersect. If they did, at the point of intersection the electric field strength, which is a vector quantity, would act in two directions at the same time.
4. A line of force must intersect an equipotential line or equipotential plane orthogonally, that is, at right angles. Otherwise, the field strength would have a finite component acting along the equipotential.
5. A line of force which terminates at a conductor must do so perpendicular to the surface. This follows from 4 above since a conducting surface under static conditions is an equipotential.

3 POTENTIAL

The word 'work' is restricted in the science of Physics to those cases in which a force has to be overcome. If a mass of 1 kg is lying on the ground it will be attracted to the earth by a force of $1 \times 9.81 = 9.81$ N. Suppose that it is lifted through a vertical distance of 1 m, then work (W) done against the gravitational force will be

$$W = Fd = 9.81 \times 1 = 9.81 \text{ N m or J}$$

This is true whatever the path through which it is lifted; if it ends at a vertical distance of 1 m from the earth's surface, the work done will be 9.81 J (figure 3.1a). The work done which, we see, is independent

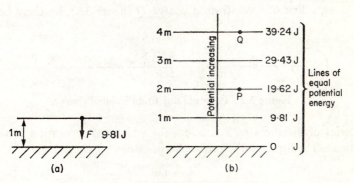

Figure 3.1   Potential and potential differences.

of the path taken is called the potential energy of the mass at the vertical height 1 m. If the mass is raised to a height of 2 m the work done will be 19.62 N m, and so on. Thus we draw a series of horizontal lines at intervals of 1 m as in figure 3.1b. The potential of a 1 kg mass at any point in line 1 will be 9.81 J; at any point on line 2 it will be 19.62 J and so on. These lines are called equipotential energy lines. The vertical distance between two points, such as P and Q will give the difference of potential energy between P and Q. Thus, in the figure it is

$$39.24 - 19.62 = 19.62 \text{ J}$$

We see that the potential of a body at any particular place is the work done in moving it from some reference position of zero potential to that place. In the gravitational case the reference position is the Earth's surface because the gravitational force is always one of attraction.

Now consider the electric field due, say, to a positively charged sphere. The probe charge q with which we investigate the properties of an electric field is always considered positive so that, unlike the gravitational case there is repulsion, not attraction. The reference position at which the potential of the probe is zero is therefore at a very great distance, theoretically infinite. Fortunately we are more concerned with differences of potential than of the absolute potential at any particular point.

Potential due to an isolated charge

Along a line of force from a charge Q (figure 3.2), let there be a

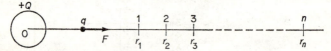

Figure 3.2   Potential due to an isolated charge.

series of distances $r_1, r_2 \ldots r_n$, for which the separation is very small. The force on the probe charge at point 1 is

$$F_1 = \frac{Qq}{kr_1^2}$$

at point 2 it is

$$F_2 = \frac{Qq}{kr_2^2} \quad \text{which is} < F_1$$

and so on. Since r_2 is very nearly equal to r_1 the mean of these distances can be taken to be the geometric mean $\sqrt{(r_1 r_2)}$, and the square of this is $r_1 r_2$. The mean force of repulsion between points 1 and 2 is therefore

$$F_{\text{mean}} = \frac{Qq}{k} \frac{1}{r_1 r_2}$$

and the work done in moving q from 2 to 1, namely the difference of potential (p.d.) between 1 and 2 is

$$\begin{aligned} W_{1.2} &= (r_2 - r_1) F_{\text{mean}} \\ &= \frac{Qq}{k} \frac{r_2 - r_1}{r_1 r_2} = \frac{Qq}{k} \left(\frac{1}{r_1} - \frac{1}{r_2} \right) \end{aligned}$$

Similarly the work done in moving q from point 3 to point 2 is

$$W_{2.2} = \frac{Qq}{k} \left(\frac{1}{r_2} - \frac{1}{r_2} \right)$$

The total work done in moving q from point n to point 1 is

$$\begin{aligned} W &= \frac{Qq}{k} \left(\frac{1}{r_1} - \frac{1}{r_2} \right) + \left(\frac{1}{r_2} - \frac{1}{r_3} \right) + \ldots \left(\frac{1}{r_{n-1}} - \frac{1}{r_n} \right) \\ &= \frac{Qq}{k} \left(\frac{1}{r_1} - \frac{1}{r_n} \right) \end{aligned}$$

If we could make q equal to unit charge (a purely mathematical convenience since the SI unit of charge is so vast that no practicable conductor could contain it), the quotient W/q is work divided by quantity. It is the potential difference $V_{1.n}$ between points 1 and n.

$$\therefore \ V_{1.n} = \frac{Q}{k} \left(\frac{1}{r_1} - \frac{1}{r_n} \right)$$

If r_n is exceedingly large so that $1/r_n$ is negligibly small we obtain the potential of point 1 due to the charge Q; it is

$$V_1 = \frac{Q}{k} \frac{1}{r_1}$$

We see that the potential at a point due to an electric charge is inversely proportional to the distance of the point from the centre of the charge. The SI units of charge and of potential difference and potential are explained in chapter 4, so that, as yet, we are not able to make numerical calculations.

B.E.—2

Field inside a charged conductor

It is stated above that a line of force starting at a charged conductor either terminates on another conductor or continues into space; it does not terminate on the conductor from which it started. We will use the concept of the tube of force. First consider two conductors A and B with A positive and B negative. Let a small area, a, of conductor A have as its share of the total quantity on A an amount Q'. Lines of force from each point of the periphery of area a will

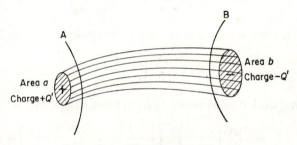

Figure 3.3   Electric tube of force.

terminate on conductor B where they will enclose another area b. The charge within this area will be equal to the charge Q', but of opposite sign, shown in figure 3.3. These peripheral lines form a tube

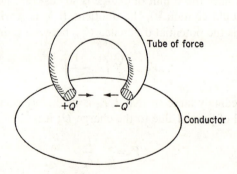

Figure 3.4   Tube of force.

of force. Now suppose that such a tube of force could start at the surface of a conductor and terminate on the same conductor as in figure 3.4. Then there would be two small charges of $+Q'$ and $-Q'$

on the same conductor. They would immediately come together and neutralise one another. Now it is obvious that if there is a line of force there must be a p.d. between its ends, from which it follows that since no line of force can both start and terminate on the same conductor, all points at its surface must be at the same potential. Hence the surface of a conductor under static conditions, that is with no current flowing, is an equipotential surface. For the same reasoning there can be no lines of force inside a conductor under static conditions.

Electric Field Strength

Suppose that a probe charge q is distant r from a charge Q, then the force on the probe charge is

$$F = \frac{Qq}{k\,r^2}$$

$$\therefore \; F/q = \frac{Q}{k\,r^2}$$

This ratio F/q is the electric field strength. If we put $q = 1$, which we can do for the purpose of investigating the unit in which electric field strength is expressed, we see that the unit is the newton/coulomb. It is shown below that the unit of electric field strength is the volt/metre, where the volt is the SI unit of potential difference, defined in chapter 4.

We have seen that the potential at a point distant d from the centre of a spherical charge, or from a point charge is

$$V = \frac{Q}{k\,r}$$

Now potential is a scalar quantity, that is, it is not directional and consequently if there is a system of separate charges Q_1, Q_2, etc. producing a resultant field at some point distant r_1, r_2, etc. from the various charges, then the resultant potential at this point is

$$V = (1/k)\,(Q_1/r_1 + Q_2/r_2 + Q_3/r_3 + \ldots)$$

Thus potential is different from electric field strength, which is a vector quantity.

Now consider a uniform electric field (figure 3.5). The equi-potential lines will be lines at right angles to the lines of force. The figure shows two such lines for potentials V_1 and V_2, V_1 being the higher potential because the lines of force reach V_1 before they reach V_2. Suppose a probe charge of $+q$ is placed in the field. If it moves from level A to level B it will have moved through a p.d. of $(V_1 - V_2)$ and therefore the work done by the field will be $(V_1 - V_2)q$.

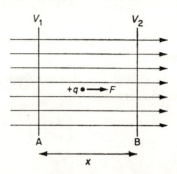

Figure 3.5   Illustrating field strength.

Let E be the strength of the field. We then define E in such a way that Eq gives the force F on the probe. If x is the distance between V_1 and V_2 then work done is Eqx,

$$\therefore\ (V_1 - V_2)\, q = Eqx$$

Thus the electric field strength is equal to the potential gradient. Now a gradient is directional and therefore E is a vector quantity.

If a charge which is free to move is situated in an electric field it will move along a line of force. If it is positive it will move in the direction of the lines of force; if negative it will move in the opposite direction. Thus in the case of the field of figure 3.5 a free positive charge will move in the direction A to B, but a free negative charge will move in the direction B to A. We say that a positive charge moves down the potential after the manner of a stone rolling downhill. A negative charge moves up the potential gradient.

Figure 3.5 is a simple energy-level diagram; this kind of diagram is of great importance in the subject of electronics.

The units in which the various entities discussed in this chapter are expressed will be explained in chapter 4.

Parallel Plates

The electric field due to two parallel plates is of great importance. Figure 3.6 shows three cases. In figure 3.6a the plates are of great area, and the field consists of parallel equally spaced equipotential energy lines, shown dashed, and equally spaced lines of force at right-angles to them. The field is uniform. Figure 3.6b shows two parallel plates of finite size. There is a law of Nature which states that any system if left to itself (that is uncontrolled by some external agency) will adjust itself to make its potential energy as small as possible.

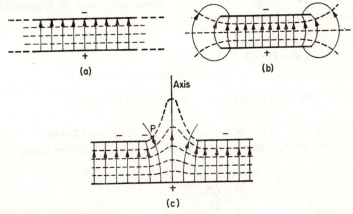

Figure 3.6 Lines of force and equipotential energy lines for the fields between parallel plates.
(a) plates of infinite area
(b) finite plates showing fringing of the flux at the edges.
(c) upper plate having a hole in the middle.

The simplest example is that of a stone of mass m raised to a height h above the ground. The work done in lifting the stone will be mgh joules and this is the magnitude of its potential energy. If released, the stone falls to the ground and its potential energy becomes zero. In the case of an electric field (and also of a magnetic field) the effect of the law is make the field take up as much room as possible. In between the plates the shape of the equipotential energy lines and of the lines of force corresponds to figure 3.6a, but outside the plates the field diverges as shown, producing what is called the fringe flux.

Figure 3.6c shows two parallel plates, one of which has a hole in the middle. There is now an extension of the electric field through the hole, the field invading the space above the hole. Suppose that a negative probe charge is placed at a point such as P. It will be acted on by a force whose direction is tangential to the line of force through P and in opposition to the direction of the line of force itself. Thus, if free to move the negative probe charge will approach the axis. A practical application of this phenomenon is given in chapter 18.

The constant k. In the SI system of units the constant k is of magnitude $1/k = 9 \times 10^9$. Suppose that two charges are equal in magnitude, then the force between them is

$$F = Q^2/(kr^2)$$
$$1/k = Fr^2/Q^2 = (Fr) \times r/Q^2$$
$$= (\text{joules} \times \text{metres})/(\text{coulomb})^2$$

The unit of $1/k$ is therefore the joule metre/coulomb².

Example 3.1. Two particles each of charge $+3.2 \times 10^{-19}$ C are 10^{-11} cm apart. Calculate the force of repulsion between them.

$$r = 10^{-13} \text{ m}$$
$$\therefore F = \frac{1}{k} \times \frac{Q^2}{r^2}$$
$$= 9 \times 10^9 \times (3.2 \times 10^{-19})^2/(10^{-13})^2$$
$$= 9.18 \times 10^{-2} \text{ N}$$

Example 3.2. A positive point charge is of magnitude 10^{-10} C. Calculate the electric field strength at distances of 1 mm and 1 cm.

$$\text{At 1 mm, } E_1 = \frac{Q}{kr^2}$$
$$= (9 \times 10^9) \times 10^{-10}/(10^{-3})^2$$
$$= 9 \times 10^5 \text{ N/C}$$
$$\text{At 10 mm, } E_2 = 9 \times 10^9 \times 10^{-10}/(10^{-2})^2$$
$$= 9 \times 10^3 \text{ N/C}$$

Example 3.3   Calculate the potentials at the above distances

$$\text{At 1 mm, } V_1 = \frac{1}{k}\frac{Q}{r_1}$$
$$= (9 \times 10^9) \times 10^{-10}/10^{-3}$$
$$= 9 \times 10^2 \text{ V/m}$$
$$\text{At 10 mm, } V_2 = (9 \times 10^9) \times 10^{-10}/10^{-2}$$
$$= 90 \text{ V/m}$$

We see that $V_1/r_1 = E_1$ and $V_2/r_2 = E_2$ showing that the potential gradient at a point is equal to the electric field strength at that point.

4 THE ELECTRIC CURRENT

Suppose that two separate points on a conductor such as a copper wire have a p.d. maintained between them; then an electric field will be set up inside the wire and it will act from the end at high potential V_1 towards the end at lower potential V_2. In this case there can be no mechanism such as that of figure 3.2 to remove this internal field because the state of the conductor is dictated by an outside agency, namely the voltage source such as a battery, which maintains the p.d. $(V_1 - V_2)$ between the ends. Consequently all the free electrons, of charge $-e$ will be acted on by a force $F = Ee$ acting in an axial direction from right to left, as shown in figure 4.1a.

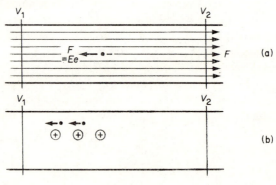

Figure 4.1

Hence there will be an axial movement of all the free electrons as in figure 4.1b in which the random motions are neglected as their net axial direction is zero. This axial movement of electrons is the electric current and we see that the carriers of electricity, namely the

32

electrons, move as a whole against the potential gradient. The accepted, or conventional, direction of an electric current is down the potential gradient, from Benjamin Franklin's choice of the terms positive and negative for the two kinds of electricity before the nature of the electric current was understood.

As explained in Chapter 1, the SI definition of the electric current is based on the forces exerted on one another by current-carrying conductors, and as it is appropriate in the present context it is given again.

The ampere is that current which, when flowing in each of two infinitely long parallel conductors in vacuo, and separated one metre between centres, causes each conductor to experience a mechanical force of 2×10^{-7} newtons per metre length.

The symbol for the unit is A; the symbol for the entity, current, is I. Thus we can write $I = 15$ A, or whatever the numeric happens to be.

In copper there are 8.5×10^{28} free electrons in each cubic metre.

The SI unit of electrical quantity

We see that an electron current is the transference in a definite direction of vast numbers of 'particles' of negative electricity called electrons. As far as we know all electrons have the same charge of $-e$ and therefore the unit of electrical quantity, or charge, is dependent on the unit of current. The SI unit is called the coulomb, and is defined as follows.

The coulomb is the quantity of electricity transported in one second by a current of one ampere.

The symbol for the unit is C and for the entity Q or q.

Experiment shows that the quantity of electricity on the electron is 1.6×10^{-19} C. Thus the number of electrons passing a unit cross section per second of a conductor carrying 1 A is

$$1/(1.6 \times 10^{-19}) = 6.25 \times 10^{18} \text{ electrons per second}$$

e.m.f. and p.d.

When resistance has to be overcome in order that motion may take place a force must be applied and work done. Thus, if water is to flow along a horizontal pipe which offers resistance to the flow it is necessary to apply a force, in this case the water pressure. Similarly when electrons are made to flow along a wire some kind of resistance is offered by the wire; for example electrons collide with the metal atoms and with one another, this impedance to motion being equivalent to a resistance. Hence, work has to be done to promote the flow of current, and in the case of a conductor in which there is no conversion of this work to some more available form of energy, the whole of the work is converted into heat and thereby dissipated. Thus the unit of heat is the same as the unit of work, namely the joule, and the unit of rate of production of heat is the watt. The si unit of electric potential, which is called the volt, is based on this dissipation of power. It is as follows:

The volt is the difference of potential between two points of a conducting wire carrying a constant current of one ampere, when the power dissipated between these points is equal to one watt.

The e.m.f. of a circuit is the total number of volts required to produce a current through the whole of the circuit, including the source. The p.d. between two points is that part of the e.m.f. which is required to send the current through that portion of the circuit included between those points. The symbol for the entity e.m.f. is E and that for the p.d. V. The symbol for the unit is V.

Resistance

The si unit of resistance is called the ohm and is defined as follows:

The ohm is the resistance between two points of a conductor when a constant difference of potential of one volt, applied between these points, produces in this conductor a current of one ampere, this conductor not being the source of any electromotive force.

The symbol for the entity is R or r and for the unit Ω.

We see from this definition that resistance is the quotient of e.m.f. and current; or, in the case of a passive part of a circuit, (that is, not containing a source of e.m.f.) of p.d. and current. Hence for a complete circuit resistance = e.m.f./current.

$$R = E/I$$
$$\therefore\ I = E/R \text{ and } E = IR$$

For part of a circuit, the part having resistance R

$$R = V/I$$
$$\therefore\ I = V/R \text{ and } V = IR$$

The above relationships are usually called Ohms' Law.

Example 4.1.   A certain metal-filament lamp has a resistance of 45Ω when cold, and 623.5Ω when hot. If the supply is at 250 V find (a) the current at the moment of switching on, (b) the current at normal working temperature.

(a) when cold $R = 45\ \Omega$
$$\therefore I = V/R = 250/45 = 5.56\ \text{A}$$
(b) when hot $R = 623.5\ \Omega$
$$\therefore I = 250/623.5 = 0.4\ \text{A}$$

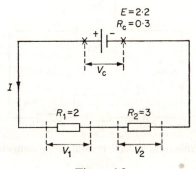

Figure 4.2

Example 4.2.   What is the power taken by the above lamp when working normally and what is the energy consumed in one hour?

$$P = VI = 250 \times 0.4 = 100\ \text{W}$$
$$W = Pt = 100 \times 60 \times 60$$
$$= 360\ 000\ \text{J}$$

Example 4.3   A certain cell has a resistance of $0.2\ \Omega$. If a resistor of $1.3\ \Omega$ is connected to its terminals a current of 1.4 A flows. What is the e.m.f. of the cell?

$$\text{Total resistance } R_t = 0.2 + 1.3 = 1.5\ \Omega$$
$$E = IR_t = 1.4 \times 1.5 = 2.1\ \text{V}$$

Example 4.4.   A circuit is made up in accordance with figure 4.2. Cell e.m.f. $= 2.2$ V, cell resistance $= 0.3\ \Omega$, $R_1 = 2.0\ \Omega$, $R_2 = 3.0\ \Omega$. Calculate the current, the p.d. across each resistor and the p.d. at the cell terminals.

Total resistance	$R_t = 0.3 + 2.0 + 3.0 = 5.3\ \Omega$
	$\therefore\ I = E/R_t = 2.2/5.3 = 0.415\ \text{A}$
p.d. across R_1	$V_1 = R_1 I = 2 \times 0.415 = 0.83\ \text{V}$
p.d. across R_2,	$V_2 = R_2 I = 3 \times 0.415 = 1.245\ \text{V}$
p.d. at cell terminals	$V_c = V_1 + V_2 = 2.075\ \text{V}$

It is very important to realise that the p.d. at the cell terminal is not $R_c I$. This is because the cell is a source of e.m.f. and the definition of the ohm states that the conductor must not be a source of e.m.f.

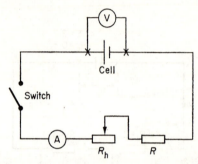

Figure 4.3   Determination of cell resistance by ammeter and voltmeter method.

Consequently the p.d. at the cell terminals is its e.m.f. less the volt-drop due to its resistance.

$$E = V_c + R_c I$$
$$\therefore\ V_c = E - R_c I$$
$$= 2.2 - 0.3 \times 0.415$$
$$= 2.075\ \text{V as before}$$

Transposing the last equation gives

$$R_c = \frac{E - V_c}{I}$$

showing that the resistance of a cell can be determined from readings of E, V_c and I. A suitable circuit is given in figure 4.3. When drawing circuit diagrams it is a good plan to draw the connections of the main circuit, i.e. that carrying the current I, by heavy lines, and the connections to voltmetres by thin lines. This is particularly advantageous with complete circuits. Also, when making up a circuit it is advisable to complete the main circuit, connecting up the voltmeters last of all.

The resistor R_h is a variable resistor enabling the current to be varied over a range appropriate to the cell under test. The first readings of the voltmeter are made with the switch open, thus giving the open-circuit e.m.f., E. Then the switch is closed and the resistance reduced in steps, the ammeter and voltmeter being read at each step.

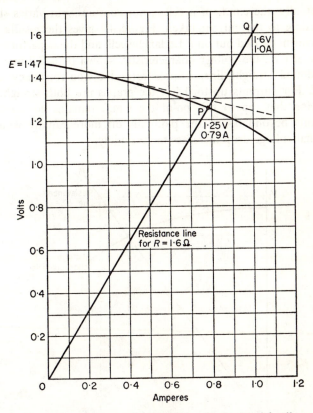

Figure 4.4 Voltage characteristic of a Leclanché cell.

With a primary cell like a Daniell cell the external resistance can be reduced to its lowest value, but if the cell is a secondary cell it is advisable to include a fixed resistor R of such value that, even with R_h completely cut out, there is no danger of a dangerously large current. With a Daniell or Leclanché cell this resistor R can be omitted.

A convenient form of tabulation is given below:

Type of cell . . .   open-circuit e.m.f. E.

Current I Terminal p.d. V_c Internal volt-drop   cell resistance
$$E - V_c \qquad R_c = \frac{E - V_c}{I}$$

Actually, the e.m.f. of a cell is not constant, but varies slightly according to the current taken. Figure 4.4 gives the results of an experiment carried out on a Leclanché cell, and the departure of the curve of p.d., the voltage characteristic, shown by a full line, from the dotted straight line, demonstrates the non-constancy of the e.m.f. For currents up to 0.4 A the graphs are both straight lines showing that, up to this point the e.m.f. remains constant. For currents beyond 0.4 A the p.d. falls off more rapidly while the internal voltage-drop increases more rapidly.

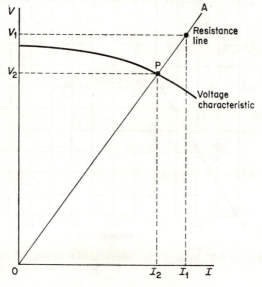

Figure 4.5   Resistance line.

Resistance Line

Figure 4.5 shows a typical voltage characteristic curve for a source of current. OA is a straight line through the origin. At any point on this line, for example point A, the p.d. of V_1 corresponds to a current of I_1. Consequently this line represents a resistance of $R_1 = V_1/I_1$ ohms. This line cuts the voltage characteristic curve at the point P for which the terminal p.d. and current are V_2 and I_2 respectively. Consequently if the cell is connected to a circuit of resistance R_1 the terminal p.d. will be V_2 and the current will be I_2.

Thus, suppose that the cell whose characteristic is given in figure 4.4 is connected to a circuit of 1.6 Ω. If such a resistance carries a current of 1 A its voltage-drop will be 1.6 V. We therefore locate the point (1.6 V, 1 A), point Q, and draw the line OQ. This is the resistance line for 1.6 Ω. It intersects the characteristic at P showing that if the cell is loaded by a resistance of 1.6 Ω, its terminal p.d. will be 1.25 V and the current will be 0.79 A.

The Inverse Square Law Again

We are now in a position to investigate the units for the entities discussed in chapter 3.

We saw that the force of repulsion, or attraction, between two small electric charges of Q_1 and Q_2 is

$$F = \frac{1}{k} \frac{Q_1 Q_2}{r^2}$$

With the SI units, the newton for F, coulomb for Q_1 and Q_2 and metre for r the constant k is equal to

$$k = 4\pi \epsilon_0$$

where ϵ_0 is a constant called the electric space constant. The numerical value of $1/(4\pi \epsilon_0)$ is 9×10^9 so that the expression for F becomes

$$F = 9 \times 10^9 \frac{Q_1 Q_2}{r^2}$$

Example 4.5.   Two positively charged particles (called α particles) each have a charge of twice that of the electron. If two such particles are 10^{-11} cm apart calculate the force of repulsion between them

$$Q_1 = Q_2 = +2e = 2 \times 1.6 \times 10^{-19} = 3.2 \times 10^{-19} \text{ C}$$
$$r = 10^{-11} \text{ cm} = 10^{-13} \text{ m}$$
$$\therefore F = 9 \times 10^9 \times \frac{(3.2 \times 10^{-19})^2}{(10^{-13})^2}$$
$$= 2.88 \times 10^{-20} \text{ N}$$

The electric strength between two planes at potentials of V_1 and V_2 volts, distant x metres apart is given by

$$E = \frac{V_1 - V_2}{x} \quad \text{V/m}$$

The unit is the volt per metre, V/m. This assumes that the field is uniform between the two planes. If the electric field is not uniform, and this is generally the case, then the field strength at any point is equal to the potential gradient at that point.

We have also seen that the force on a charge Q placed in an electric field of strength E is

$$F = EQ$$
$$\therefore \quad E = F/Q$$

The unit of force is the newton and the unit of charge the coulomb, showing that the unit of electric field strength is also the newton per coulomb, N/C.

Example 4.6.   Two parallel plates are 1 cm apart and a p.d. of 200 V exists between them. What is the force acting on an electron in the space between the plates and what is its direction?

$$E = 200/10^{-2} = 2 \times 10^4 \text{ V/m or N/C}$$
$$Q = e = 1.6 \times 10^{-19} \text{ C}$$
$$\therefore \quad F = 2 \times 10^4 \times 1.6 \times 10^{-19}$$
$$= 3.2 \times 10^{-15} \text{ N}$$

The direction of the force is from negative to positive plate along a line perpendicular to the planes of the plates.

Example 4.7.   Given that the mass of the electron is 9.1×10^{-31} kg, what is the acceleration of the electron in example 4.6?

$$F = m a$$
$$\therefore \quad a = F/m = \frac{3.2 \times 10^{-15}}{9.1 \times 10^{-31}} = 3.5 \times 10^{15} \text{ m/s}^2$$

The beginner may find some of the electrical units difficult to understand. There could be no physical science without an adequate system of units and therefore there can be no understanding of a physical science without a knowledge of its units.

Summary of Units

Name of entity	Symbol for entity	Name of SI unit	Symbol for unit
Electric current	I	ampere	A
E.m.f. and p.d.	E, V	volt	V
Resistance	R	ohm	Ω
Quantity or charge	Q	coulomb	C
Electric field strength	E	volt per metre or newton per coulomb	V/m or N/C

5 POWER, ENERGY, HEAT

The number of amperes flowing in a circuit = the number of coulombs /second.

The number of watts = the product of the numbers of volts and amperes.

The number of joules = the product of the numbers of watts and seconds

$$= \text{volts} \times (\text{amperes} \times \text{seconds})$$
$$= \text{volts} \times \text{coulombs}$$

Hence J $\quad = VQ$ or VIt

Example 5.1. A secondary cell of 2.2 V delivers a current of 10 A for half an hour to an external circuit. The resistance of the cell is 0.04 Ω. Calculate the energy expended (a) in the whole circuit, (b) in the external circuit.

$$\text{Let } R_c = \text{cell resistance}$$

Then internal volt-drop

$$R_c I = 0.04 \times 10 = 0.4 \text{ V}$$
$$\therefore \text{ Terminal p.d. } \quad V = 2.2 - 0.4 = 1.8 \text{ V}$$

For the whole circuit the energy expended

$$W = EIt = 2.2 \times 10 \times (30 \times 60)$$
$$= 39\ 600 \text{ J}$$

For the external circuit

$$W = VIt = 1.8 \times 10 \times (30 \times 60)$$
$$= 32\ 400 \text{ J}$$

The following relationships should be memorised

1. Power | Complete circuit | Part of a circuit

	Complete circuit	Part of a circuit	
$P =$	EI	VI	
$P =$	$E \times E/R$	$V \times V/R$	
$=$	E^2/R	V^2/R	watts
$P =$	$R_t I \times I$	$RI \times I$	
$=$	$I^2 R_t$	$I^2 R$	

The left-hand set refers to the complete circuit, and the right-hand set to that portion of the circuit which is of such resistance R that the voltage drop along it is V.

2. Energy expended

$$W = EIt \qquad \text{or} \qquad VIt$$
$$W = (E^2/R)\, t \qquad \text{or} \qquad (V^2/R)\, t \qquad \text{joules}$$
$$W = I^2 R\, t \qquad \text{or} \qquad I^2 Rt$$

Example 5.2. A battery of e.m.f. 100 V has a resistance of 5 Ω. It is connected to an external circuit of resistance 15 Ω. What will be the power intake of this 15 Ω resistor?

$$R_t = R_c + R$$
$$= 5 + 15 = 20 \ \Omega$$
$$\therefore \ I = E/R_t = 100/20 = 5 \ \text{A}$$

Voltage drop in external resistor
$$V = IR = 5 \times 15 = 75 \ \text{V}$$

∴ Power intake of resistor
$$P = VI = 75 \times 5 = 375 \ \text{W}$$

or, alternatively
$$P = I^2 R = 5^2 \times 15 = 375 \ \text{W, as before.}$$

Example 5.3.   A battery of e.m.f. 10 V delivers a steady current of 5 A for 10 min. What will be the energy expended in the circuit, and the total quantity of electricity passed round the circuit?

$$P = EI \text{ for the whole circuit}$$
$$= 10 \times 5 = 50 \text{ W}$$
$$\therefore \quad W = Pt \quad = 50 \times (10 \times 60) = 30\ 000 \text{ J}$$
$$Q = It \quad = 5 \times 10 \times 60 \quad = \quad 3000 \text{ C}$$

Heat production

When resistance is overcome and the energy expended to accomplish this is not converted into some other form of energy, it is converted into heat. There are many everyday examples, such as the rise in temperature of the brakes of a car when they are applied, the rise in temperature of the bearings of an engine, and so on. Similarly, in an electric circuit, if the applied p.d. has nothing to do but overcome the resistance of the circuit, then the whole of the energy expended will be converted into heat energy. If the circuit to which the p.d. is applied is not passive, but contains a source of e.m.f. such as a battery, then only that portion of the energy concerned with the overcoming of resistance will be converted into heat.

Thus if a current I flows through a resistor R for t seconds the energy converted into heat is

$$W = I^2 R t \text{ joules}$$

Note that (i) the heat generated is proportional to the square of the current, (ii) it is proportional to the resistance, (iii) it is proportional to the time the current is passing.

If electrical energy is converted into heat energy, as in an electric kettle, then 4180 joules are required to raise the temperature of 1 kg of water by 1° C or 1 K (the temperature *differences* in the two scales being equal). This amount of heat energy is called the kilocalorie (kcal) and therefore 1 kcal = 4180 J. Hence, if all the electrical energy is converted into heat energy, the heat generated

$$H \text{ (kcal)} = W \text{ (joules)}/4180.$$

This value of 4190 is called the mechanical equivalent of heat symbol J, and was first determined by the English physicist Joule in 1843. The name is derived from the fact that in his experiments mechanical energy, not electrical energy, was converted into heat energy.

The heat energy of a mass M kg of water at a temperature of $\theta°$ C is $M\theta$ kcal. The heat energy of any other substance is $MS\theta$ where S is the specific heat of that substance. For example, the specific heat of copper is 0.094.

Example 5.4.   An electric heater raises the temperature of 10 kg of water from 15° C to boiling point. 75 per cent of the energy intake is actually utilised in heating the water, the rest is lost by radiation. Calculate the energy intake of the heater

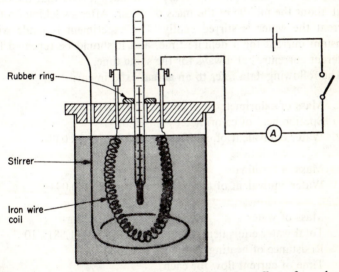

Figure 5.1   Experiment to demonstrate the heating effect of an electric current.

Heat energy of the water content
$$= 10 \times (100 - 15) = 850 \text{ kcal}$$
∴ Energy intake of the heater in heat units
$$H = 850/0.75 = 1133 \text{ kcal}$$
∴ Energy intake in joules
$$W = 1133 \times 4180$$
$$= 4.75 \times 10^6 \text{ J}$$

The following experiment demonstrates the above relationships. A thin wire of some high resistance material, such as manganin is wound into a close spiral and its ends soldered to two brass or copper rods; connectors are screwed to the rods so that the coil can be connected to a source of current, such as a 2 V secondary cell. A series circuit is made up of cell, coil, controlling rheostat, ammeter and switch. The coil is immersed in water in a calorimeter, as shown in figure 5.1. The mass of water is determined by weighing the calorimeter first empty and then containing water. To prevent loss of heat from radiation the calorimeter can stand on a pad of thick felt, and also be enclosed in a felt jacket. The calorimeter can also be closed by a wooden lid which holds various items of the apparatus, as shown. The thermometer should be at such a level that the bulb is at about the middle of the mass of water. After switching on the current the water is stirred gently. The experiment is made with constant current for a definite time, and it should be repeated for different currents, but always for the same time.

The following data refer to an actual experiment:

$$
\begin{aligned}
\text{Mass of calorimeter} &= 83.5 \times 10^{-3} \text{ kg} \\
\text{Specific heat of calorimeter} &= 0.094 \\
\therefore \text{ Water equivalent of calorimeter} &= 83.5 \times 10^{-3} \times 0.094 \\
&= 7.8 \times 10^{-3} \text{ kg} \\
\text{Mass of terminals} &= 20 \times 10^{-3} \text{ kg} \\
\therefore \text{ Water equivalent of terminals} &= 20 \times 10^{-3} \times 0.094 \\
&= 1.88 \times 10^{-3} \text{ kg} \\
\text{Mass of water} &= 250 \times 10^{-3} \text{ kg} \\
\therefore \text{ Total water equivalent, } m &= (250 + 7.8 + 1.88) \times 10^{-3} \\
\text{Resistance of heating coil} &= 3.25 \text{ } \Omega \\
\text{Time of current flow for each} & \\
\text{value of the current} &= 15 \text{ min}
\end{aligned}
$$

Current (I)	Initial temp. (θ_1)	Final temp. (θ_2)	$(\theta_2 - \theta_1)$	H, kilocal $= m/(\theta_2 - \theta_1)$	I^2
0.5	19.8	20.5	0.7	181×10^{-3}	0.25
0.75	20.5	21.9	1.4	363×10^{-3}	0.56
1.00	22.0	24.7	2.7	701×10^{-3}	1.00
1.25	16.0	20.1	4.1	1065×10^{-3}	1.56
1.50	20.1	26.5	6.4	1665×10^{-3}	2.25
1.75	27.1	35.0	8.0	2080×10^{-3}	3.03
2.00	35.0	45.4	10.4	2700×10^{-3}	4.00

The heat generated, as given in column 5, is plotted against the square of the current in figure 5.2. Allowing for experimental errors, which are inevitable unless elaborate precautions are taken, and corrections made, it will be seen that the experimental points lie on a straight line, showing that the heat generated in a specific time is proportional to the square of the current. This assumes a constant

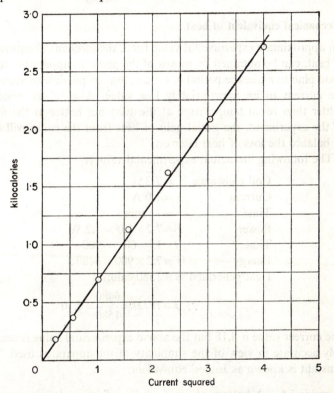

Figure 5.2   Result of the experiment, demonstrating that the heat production in a given time is proportional to the square of the current.

resistance for the heating coil; this is the case when manganin is used. Iron wire is not suitable because its resistance increases appreciably with temperature rise.

Example 5.5   A battery of e.m.f. 20 V and internal resistance 1 Ω supplies a resistor of 5 Ω. How many kilocalories will be generated in the resistor in 10 minutes?

$$R_t = 1\ \Omega + 5\ \Omega = 6\ \Omega$$
$$I = E/R_t \quad = 20/6 = 3.33\ \text{A}$$

Power consumed in resistor $P = (3.33)^2 \times 5 = 55.6\ \text{W}$

$\therefore$ In 10 minutes $W = 55.6 \times 600 = 33\ 360\ \text{J}$

$$\therefore\ H = \frac{33\ 360}{4180} = 7.97\ \text{kcal}$$

Mechanical equivalent of heat

An approximate experimental value for J, the mechanical equivalent of heat, can be obtained by means of the previous apparatus. It is a good plan to make the period of time as long as possible by reducing the current to an appropriately low value. Also, if the water is colder than room temperature at the start but hotter at the finish of the experiment, the initial gain of heat from the room will tend to balance the loss of heat later on.

The following are actual experimental values:

Coil resistance	$= 7.2\ \Omega$
Current	$= 1.0\ \text{A}$
Time	$= 15\ \text{min}$
Power	$P = 7.2 \times 1.0^2 = 7.2\ \text{W}$
Time	$t = 15 \times 60\ \ = 900\ \text{s}$
Energy	$W = 7.2 \times 900 = 6480\ \text{J}$
Heat generated	$H = 1580\ \text{calories}$

$$\therefore\ J = W/H = \frac{6480}{1580} = 4.10$$

The correct value is 4.18 but the above experimental value is reasonably accurate in view of the simplicity of the apparatus used. This constant is known as Joules' equivalent.

Example 5.6.   A battery of e.m.f. 20 V and internal resistance 1 Ω supplies a resistor of 5 Ω. How many calories will be generated in the resistor in 10 minutes?

$$R_t = 1\ \Omega + 5\ \Omega = 6\ \Omega$$
$$\therefore\ I = E/R \qquad = 20/6 = 3.33\ \text{A}$$

Power dissipated in resistor $R = 3.33^2 \times 5 \quad = 55.6\ \text{W}$

$$t = 10 \times 60 \qquad = 600\ \text{s}$$

$\therefore$ Energy

$$W = 55.6 \times 600 = 33\ 360\ \text{J}$$
$$H = W/J \qquad = 33\ 360/4.18$$
$$= 7970\ \text{kcal}$$

Example 5.7.   In a certain box of resistors each coil is capable of radiating 3 W. What is the highest voltage that can be safely applied to the 200 Ω coil, and how much heat will be dissipated if the coil is left in circuit for half an hour?

$$\text{coil watts} = V^2/R$$
$$\therefore \ 3 = V^2/200$$
$$V^2 = 600, \quad V = 24.5 \text{ V}$$
$$\therefore \ W = 3 \times 30 \times 60 \text{ J}$$
$$H = W/J$$
$$= (3 \times 30 \times 60)/4180 = 1.290 \text{ kcal}$$

Cost of electrical energy

The joule is a very small unit, and for commercial purposes it is necessary to use a unit of more convenient magnitude. The unit is the kilowatt-hour, kWh. It is equal to the expenditure of power of 1 kilowatt for one hour.

$$\therefore \ 1 \text{ kWh} = 1000 \times 60 \times 60 = 3\ 600\ 000 \text{ J}$$

Example 5.8.   An electric kettle holds 1.7 kg of water. The required change in temperature to bring the water to boiling point is 85 K. If the heating element takes 500 W, how long will it take? If the efficiency of the kettle is 85 per cent and one kWh costs 1p, assuming a flat rate, what will be the cost?

$$M = 1700 \text{ g}$$
$$= 1.7 \text{ kg}$$
$$\text{Temp. rise} = 85 \text{ K} \quad = 85° \text{ C}$$
$$\therefore \ H = 1.7 \times 85 \ = 144.5 \text{ kcal}$$
$$\therefore \ W = 4180 \ H \ = 4180 \times 144.5$$
$$= 602\ 000 \text{ J}$$

There are 1000 W to the kW and 3600 seconds to the hour

$$\therefore \ \text{kWh utilised} = \text{joules}/(1000 \times 3600)$$
$$= 602\ 000/3\ 600\ 000$$
$$= 0.166 \text{ kWh}$$

This would be the energy taken from the supply if there were no heat loss from the kettle. But the kettle is only 85 per cent efficient.

$$\therefore \quad \text{kWh consumed} = 0.166/0.85 = 0.196$$
$$\therefore \quad \text{Cost} \qquad\qquad = 0.196 \times 1 \quad = 0.196\text{p}$$
$$\text{Time taken} \qquad = (\text{kWh/kW}) \text{ hours}$$
$$\text{and kW} \qquad\quad = 500/1000 = 0.5$$
$$\therefore \quad \text{Time taken} \quad = 0.196/0.5 = 0.392 \text{ hour}$$

Example 5.9. The motor driving a mine fan has an output of 225 kW when the fan is running at full speed. Its efficiency is then 92 per cent. Calculate the cost per year of running the fan continuously at full speed if the cost of one kWh at the mine is 0.5p.

$$\text{Motor intake } P = 225/0.92$$
$$= 244.5 \text{ kW}$$
$$\text{No. of hours per annum of 365 days}$$
$$= 365 \times 24$$
$$\therefore \quad \text{kWh per annum} = 244.5 \times 365 \times 24$$
$$= 2.14 \times 10^6$$
$$\tfrac{1}{2}\text{ p} = \pounds(1/200)$$
$$\therefore \quad \text{Annual cost} \quad = \frac{2.14 \times 10^6}{200}$$
$$= \pounds10\ 700$$

6 ELECTRIC CIRCUITS

There are two basic ways in which electric circuits can be made up, namely the series arrangement and the parallel arrangement.

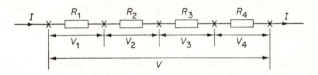

Figure 6.1   Resistors in parallel.

Figure 6.1 shows four resistors connected in series. There is only one path for the current. It is obvious that the total resistance must be equal to the sum of the separate resistances; the proof is as follows. The figure shows only a part of a circuit and therefore the voltage V between the ends is the total voltage-drop, not the e.m.f. This must be equal to the sum of the individual voltage-drops V_1, V_2 etc.

$$\therefore\ V = V_1 + V_2 + V_3 + V_4$$

This is not a law of universal application since, in general, it only holds when the current is constant in magnitude, that is does not change from instant to instant. We are only concerned with steady currents and so the equation applies. Applying Ohm's law to the complete circuit, calling its resistance R, we have $V = IR$. Applying it to each resistor in turn we have

$$V_1 = IR_1,\ V_2 = IR_2,\ \text{etc.}$$
$$\therefore\ IR = IR_1 + IR_2 + IR_3 + IR_4$$
$$R = R_1 + R_2 + R_3 + R_4$$

In general, if there are n resistors in series

$$R = R_1 + R_2 + \ldots + R_n$$
$$\text{or } R = \Sigma_1^n (R)$$

In words, for resistors in series the total resistance is equal to the arithmetic sum of the separate resistances.

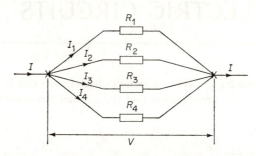

Figure 6.2

Figure 6.2 shows four resistors connected in parallel. We see that they form separate branches of a circuit and that the total current from the rest of the circuit is divided between them. Since an electric current is due to the axial movement of electrons along conductors it follows that the number of electrons leaving the left-hand junction must be equal to the number reaching the right-hand junction.

$$\therefore \ I = I_1 + I_2 + I_3 + I_4$$

Let V be the p.d. across the junction, then it is also the p.d. across each separate resistor. Denote the single resistor which is equivalent to the branched arrangement by R, then since $I = V/R$,

$$\frac{V}{R} = \frac{V}{R_1} + \frac{V}{R_2} + \frac{V}{R_3} + \frac{V}{R_4}$$

$$\frac{1}{R} = \frac{1}{R_1} + \frac{1}{R_2} + \frac{1}{R_3} + \frac{1}{R_4}$$

Or, in general, with n resistors in parallel

$$\frac{1}{R} = \frac{1}{R_1} + \frac{1}{R_2} + \ldots + \frac{1}{R_n}$$

$$\frac{1}{R} = \Sigma_1^n \left(\frac{1}{R}\right)$$

In words, the reciprocal of the total resistance is equal to the sum of the reciprocals of the branch resistances.

The current in each branch in terms of the total current is determined as follows:

$$V = IR = I_1 R_1$$
$$\therefore \; I_1 = (R/R_1)I$$

and similarly with I_2, I_2 etc.

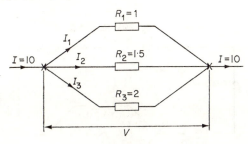

Figure 6.3

Example 6.1.   Three resistors of 1, 1.5 and 2 Ω respectively are in parallel and they share a current of 10 A. Find the p.d. across them and the current in each resistor (figure 6.3).

$$1/R = 1/R_1 + 1/R_2 + 1/R_3$$
$$= 1/1 + 1/1.5 + 1/2 \; = 2.17$$
$$\therefore \quad R = 1/2.17 = 0.462 \; \Omega$$
$$V = IR \quad = 10 \times 0.462 = 4.62 \text{ V}$$
$$I_1 = V/R_1 \; = 4.62/1 \qquad = 4.62 \text{ A}$$
$$I_2 = V/R_2 = 4.62/1.5 \; = 3.07 \text{ A}$$
$$I_3 = V/R_3 = 4.62/2 \qquad = 2.31 \text{ A}$$
$$\text{Sum} \qquad = \underline{10.0 \text{ A}}$$

alternatively

$$I_1 = (R/R_1)I = (0.462/1) \times 10 \qquad = 4.62 \text{ A}$$
$$I_2 = (R/R_2)I = (0.462/1.5) \times 10 = 3.07 \text{ A}$$
$$I_3 = (R/R_3)I = (0.462/2) \times 10 \; = 2.31 \text{ A as before}$$

Example 6.2   To obtain the resistance of an appliance of unknown value it is connected in parallel with a second resistor of 2.5 Ω. The two are connected to a battery and the p.d. across them and the total current measured. These are 2 V and 3.5 A respectively. What is the unknown resistance?

Use subscripts x and 2 for the two resistors.

$$I_2 = 2.0/2.5 = 0.8 \text{ A}$$
$$\therefore \quad I_x = 3.5 - 0.8 = 2.7 \text{ A}$$
$$\therefore \quad R_x = V/I_x = 2/2.7 = 0.74 \text{ } \Omega$$

In most practical circuits series and parallel connections exist together.

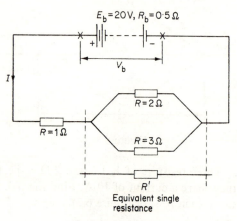

Figure 6.4

Example 6.3.   A battery of e.m.f. 20 V and internal resistance 0.5 Ω is connected to a circuit consisting of a resistor of 1 Ω in series with two resistors of 2 Ω and 3 Ω in parallel. Calculate the total current, the two branch currents, the battery terminal p.d. and the volt drop along each resistor.

The circuit is given in figure 6.4. Dealing with the branched portion first and using the suffixes shown in the figure, the resistance R',

$$R' = \frac{R_2 R_3}{R_2 + R_3} = \frac{2 \times 3}{5} = 1.20 \text{ } \Omega$$
$$\therefore \quad R_t = 0.5 + 1 + 1.2 = 2.7 \text{ } \Omega$$
$$\therefore \quad I = E/R_t = 20/2.7 = 7.405 \text{ A}$$

Voltage drop in battery $R_t\,I = 0.5 \times 7.405 = 3.703$ V

$$\therefore \quad \text{Battery p.d. } V_b = 20 - 3.703 = 16.297 \text{ V}$$

Voltage drop along R_1, $R_1 I = 1 \times 7.405 = 7.405$ V

Voltage drop along branched part of circuit

$$R' I = 1.2 \times 7.405 = 8.89 \text{ V}$$

$$I_1 = \frac{R' I}{R_1} = \frac{8.89}{2}$$
$$= 4.445 \text{ A}$$

$$I_2 = \frac{R'I}{R_2} = \frac{8.89}{3}$$
$$= 2.96 \text{ A}$$

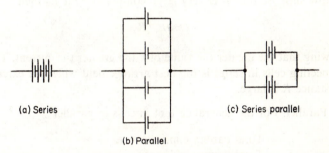

(a) Series   (b) Parallel   (c) Series parallel

Figure 6.5   Alternative methods of connecting cells to form a battery.

Note that the sum of the various voltage drops which take place external to the battery must be the same as the battery terminal p.d. because it is this p.d. which is utilised externally.

$$V_1 = 7.405 \times 1 = 7.405 \text{ V}$$
$$V_2 = 7.405 \times 1.2 = 8.890 \text{ V}$$
$$\text{Sum} = 16.295 \text{ V}$$

Arrangement of cells

Like resistors, cells can be connected in series or in parallel, figure 6.5a and b. They can also be connected in series–parallel, figure 6.5c.

Let E_c = e.m.f. per cell
R_c = resistance per cell
R = external resistance

56 BASIC ELECTROTECHNOLOGY

(a) Series. For the general case of n cells in a row

$$\text{Total e.m.f.} \quad E = nE_c$$
$$\text{Total battery resistance} = nR_c$$
$$\text{Total circuit resistance} = nR_c + R$$
$$\therefore I = \frac{nE_c}{nR_c + R}$$

Suppose R is very small then

$$I \simeq nE_c/nR_c = E/R_c$$

resistance putting a number of cells in series is little better than using only one cell.

Now suppose that R is very large compared with R_c then

$$I \simeq nE_c/R$$

showing that the greater the value of n the greater the current. Thus connecting cells in series is the best arrangement when the external resistance R is large.

(b) Parallel. For the general case of m cells in parallel

$$\text{Total battery e.m.f.} \quad = E_c$$
$$\text{Total battery resistance} = R_c/m$$
$$\text{Total circuit resistance} = (R_c/m) + R$$
$$\therefore I = \frac{E_c}{(R_c/m) + R}$$
$$= \frac{mE_c}{R_c + mR}$$

Suppose that R is very small then

$$I \simeq mE_c/R_c$$

showing that, the greater the value of m the greater the current.

Now suppose that R is very large compared with R_c, then

$$I \simeq \frac{mE_c}{mR} = E_c/R$$

which shows that, in such a case, connecting a number of cells in parallel is little better than using only one cell. Thus connecting

cells in parallel is best when the external resistance is very small.
(c) Series–Parallel (figure 6.5c). Let there be m rows in parallel each
having n cells in series.

$$\text{e.m.f. per row} = nE_c$$
$$\text{Resistance per row} = nR_c$$
$$\therefore \text{Resistance per } m \text{ rows} = (n/m)\, R_c$$
$$\therefore \text{Circuit resistance} = (n/m)\, R_c + R$$
$$I = \frac{nE_c}{(n/m)\, R_c + R}$$
$$= \frac{mnE_c}{nR_c + mR}$$

Example 6.4.   A battery consists of 6 cells each of e.m.f. 1.5 V and
resistance 2 Ω. The external resistance is 5 Ω. Calculate the possible
values of the current.

(a) Simple series
$$E = 6 \times 1.5 = 9 \text{ V}$$
$$R_b = nR_c = 6 \times 2 = 12 \ \Omega$$
$$R_t = 12 + 5 = 17 \ \Omega$$
$$\therefore I = 9/17 = 0.529 \text{ A}$$

(b) Simple parallel
$$E = E_c = 1.5 \text{ V}$$
$$R_b = R_c/m = 2/6 = 0.33 \ \Omega$$
$$R = 0.33 + 5 = 5.33 \ \Omega$$
$$I = 1.5/5.33 = 0.281 \text{ A}$$

(c) Series–parallel
We can have two rows of three cells or three rows of two cells.
With $m = 2$ and $n = 3$, $nR_c = 6$ and $mR = 10$

$$\therefore I = \frac{6 \times 1.5}{6 + 10}$$
$$= 9/16 = 0.56 \text{ A}$$

With $m = 3$ and $n = 2$, $nR_c = 4$ and $mR = 15$

$$\therefore I = \frac{6 \times 1.5}{4 + 15}$$
$$= 9/19 = 0.47 \text{ A}$$

It can be shown that the current is a maximum when the external resistance is as nearly equal to the battery resistance as possible. Thus in the above example when $m=2$ the two values in the denominator, 6 and 10, are more nearly equal than when $m=3$, for which the two values in the denominator are 4 and 15.

7 ELECTRICAL RESISTANCE

It will be obvious that if a wire is of uniform cross-section and its material is homogeneous, its electrical resistance will be proportional to its length l.

$$\therefore \ R \propto l$$

Now suppose that there are two conductors which are identical as to material, length and cross-section, and that they are joined in parallel. Let a p.d. V be applied and let the current be I, then the resistance of the two wires in parallel is

$$R_1 = V/I$$

Each wire alone will have a current of $I/2$ so that for each wire

$$R_2 = 2V/I = 2R_1$$

But the only difference between the two cases is that with two conductors in parallel the cross-section is doubled. Thus doubling the cross-section halves the resistance, and similarly for any other ratio of the two cross-sections. Thus, in general, the resistance is inversely to the cross-section.

$$R \propto (1/a)$$

Combining the two, we have

$$R \propto l/a$$

We require an equation, not an expression of proportionality and therefore we introduce a constant ρ, giving

$$R = \rho l/a$$

copper, silver, etc.; its physical state, such as temperature; and also on the system of units. If we put $l=1$ m and $a=1$ m² in the above equation we have $R=\rho$, showing that ρ is the resistance of a unit cube of the material, as shown in figure 7.1. It is called the resistivity. Thus the resistivity of a conductor is the resistance of a cube of the material of the conductor whose edges are of 1 m length. The name of the unit is the ohm-metre. It is sometimes called the ohm per metre cube but this can lead to confusion and should not be used.

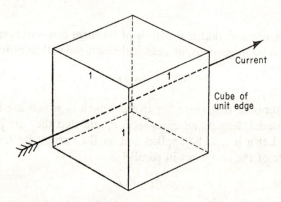

Figure 7.1   A conductor in the form of a unit cube.

Example 7.1.   Two thousand metres of wire of 0.1 cm diameter has a resistance of 20 Ω. What is the resistance of ten thousand metres of wire of the same material but of 0.2 cm diameter?

$$R_1 \propto l_1/a_1 \propto l_1/d_1^2$$
$$\text{and } R_2 \propto l_2/a_2 \propto l_2/d_2^2$$

$$\therefore \quad \frac{R_2}{R_1} = \frac{l_2}{d_2^2} \times \frac{d_1^2}{l_1}$$
$$= (l_2/l_1) \times (d_1/d_2)^2$$
$$= (10\,000/2000) \times (0.1/0.2)^2 = 1.25$$
$$\therefore \quad R_2 = 1.25 \times 20 = 25 \text{ Ω}$$

Example 7.2.   One hundred metres of conductor 1 of circular cross-section and diameter 0.048 cm has a resistance of 1.35 of conductor 2 of square cross-section has a resista the resistivity of conductor 2 is 1.7 times that of conductor 1 what is the size of the wire.

$$R_1 = \frac{\rho_1 l_1}{a_1} ; \quad R_2 = \frac{\rho_2 l_2}{a_2}$$

$$\therefore \quad \frac{R_1}{R_2} = \frac{\rho_1}{\rho_2} \times \frac{l_1}{l_2} \times \frac{a_2}{a_1}$$

$$\therefore \quad a_2 = \frac{R_1}{R_2} \times \frac{\rho_2}{\rho_1} \times \frac{l_2}{l_1} \times a_1$$

$$= \frac{1.35}{2.5} \times \frac{1.7}{1} \times \frac{50}{100} \times \left(\frac{\pi}{4} \times 0.048^2 \right)$$

$$= 0.361 \times (0.048)^2$$

$\therefore$ Side of square section

$$= (0.361)^{\frac{1}{2}} \times 0.048$$

$$= 0.029 \text{ cm}$$

Example 7.3. Calculate the resistance of 150 cm of copper wire of 1 mm diameter given that $\rho = 1.72 \times 10^{-8}$ ohm-metre.

$$a = \pi r^2 = \pi \times (0.5 \times 10^{-3})^2 = \pi \times 25 \times 10^{-8} \text{ m}^2$$

$$l = 150 \text{ cm} = 1.5 \text{ m}$$

$$\therefore \quad R = \frac{\rho l}{a} = \frac{1.72 \times 10^{-8} \times 1.5}{\pi \times 25 \times 10^{-8}}$$

$$= 0.0385 \ \Omega$$

The value of the resistivity of any particular conductor depends on the temperature and therefore, in any table of resistivities the temperature to which the values refer must be stated. The following table will be found useful. Copper is the most widely used electrical conductor, being used almost exclusively for the winding of all kinds of electrical machines and for the conducting cores of most electrical wires and cables. Aluminium is used largely for the conductors of high-voltage overhead transmission lines, but also for the cores of cables and, in a few special cases, the windings of electrical machines. German silver and manganin are, as the table shows, metals of high resistivity, and are therefore used for the coils of resistance boxes and other apparatus used for accurate measurement. Iron is of medium resistance and is cheap; it is often used for the controlling resistances for electric motors.

Table of Resistivities at 293 K (20° C)

Material	Resistivity in Ω m	Mean temperature coefficient over a range of 100 K
Aluminium	2.82×10^{-8}	0.004
Copper	1.72×10^{-8}	0.004
Iron	9.8×10^{-8}	0.006 (about)
Constantin	48×10^{-8}	0
German silver	30×10^{-8}	0.00025
Manganin	44×10^{-8}	0

Variation of Resistance with Temperature

Most conductors experience an increase in electrical resistance when the temperature is increased. Imagine again the movement of electrons along a wire when a p.d. is applied to the ends, and consider the opposition to this motion caused by repeated collisions with the ions. We have so far regarded the ions as occupying fixed positions, but, in fact, they are not rigidly fixed but can oscillate to a certain extent about a mean position. This oscillation is a measure of the temperature of the metal, and the higher the temperature the greater will be the amplitude of the oscillations.

Figure 7.2 shows in a greatly simplified manner, with the random thermal motions of the electrons omitted, that the chance of a journey without collision, even over a very short distance becomes less and less as the amplitude of the ionic oscillation increases. The oscillating ions deflect the electrons and thereby cause an increase in resistance.

If the temperature is reduced the amplitude of the ionic oscillation decreases and the chance of collision therefore decreases. Experiments carried out at temperatures only a few degrees above absolute zero have proved that, with certain metals, there is a tendency to zero resistance at such temperatures. The phenomenon is called superconductivity. Thus lead and tin lose all traces of electrical resistivity at about 4 K. This phenomenon is by no means as simple as stated above, and is not yet completely understood.

Imagine then a conductor which has a resistance of 1 Ω at some reference temperature, say 273 K, or, more conveniently, 0° C. Let an increase in temperature of one degree cause the resistance to increase by a small amount a; then, assuming a linear relationship, an increase in temperature of 2 degrees will cause an increase in resistance of $2a$, and so on.

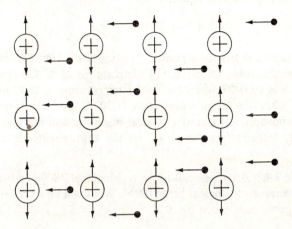

Figure 7.2   Effect of temperature on the chance of collisions of electrons and ions.

$$\text{Resistance at } 0°\text{ C} = 1 \ \Omega$$
$$\text{Resistance at } 1°\text{ C} = (1+a) \ \Omega$$
$$\text{Resistance at } 2°\text{ C} = (1+2a) \ \Omega$$
$$\text{etc.} \qquad\qquad \text{etc.}$$
$$\text{Resistance at } t°\text{ C} = (1+at) \ \Omega$$

If we had started with a conductor of initial resistance R_0 at $0°$ C, the resistance at any temperature t would be

$$R = R_0 \,(1 + at)$$

The term a is the temperature coefficient of resistance.

In practice it will be very rare for a conductor to start at $0°$ C, its initial temperature probably being that of the surrounding atmosphere. Calling the initial temperature t_1 and the final temperature t_2 we have

$$R_1 = R_0 \,(1 + at_1)$$
$$R_2 = R_0 \,(1 + at_2)$$
$$\therefore \ \frac{R_2}{R_1} = \frac{1 + at_2}{1 + at_1}$$

If α is small this is very nearly equal to

$$1 + \alpha \, (t_2 - t_1)$$

giving the approximate equation

$$R_2 = R_1 \, [1 + \alpha \, (t_2 - t_1)]$$

This equation is strictly accurate if we use for α, not the value at $0°$ C but the value at the initial temperature of $t_1°$ C. Thus with copper α is 0.00427 when the initial temperature is $0°$ C but it is 0.00378 when the initial temperature is $30°$ C. For the temperature ranges appropriate to most electrical machines and apparatus it is generally sufficiently accurate to use the approximate equation.

Example 7.4.   A copper coil has a resistance of 1000 Ω when cold, the temperature coefficient being 0.004. What will be its resistance if the temperature is raised by 20 K

$$R_1 = 1000 \ \Omega$$
$$t_2 - t_1 = 20 \text{ K (or } 20° \text{ C)}$$
$$\therefore \ R_2 = 1000 \, (1 + 0.004 \times 20)$$
$$= 1080 \ \Omega$$

Transposing the simplified equation gives

$$\alpha \, (t_2 - t_1) = (R_2/R_1) - 1$$
$$= \frac{R_2 - R_1}{R_1}$$
$$\therefore \ t_2 - t_1 = \frac{R_2 - R_1}{R_1 \, \alpha}$$

Thus, if R_1 and R_2 are measured the temperature rise can be calculated. This is a very important method of determining the temperature of coils which are heated by the passage of current. A thermometer in contact with the outside gives merely the surface temperature, whereas the change of resistance method gives the mean temperature.

Example 7.5.   A magnetising coil of a certain machine takes a current of 2.38 A at 500 V when the machine is cold at 293 K. When the machine has warmed up the current is 2.17 A. Calculate the average temperature of the winding given that a for the initial temperature of 293 K a is 0.00393

$$R_1 = V/I_1 = 500/2.38 = 210 \ \Omega$$
$$R_2 = V/I_2 = 500/2.17 = 230 \ \Omega$$
$$\therefore \quad t_2 - t_1 = (R_2 - R_1)/R_1 \ a$$
$$= (230 - 210)/(210 \times 0.00393)$$
$$= 24.3°$$
$$\therefore \quad t_2 = 293 + 24.3 = 317.3 \ \text{K} = 44.3° \ \text{C}$$

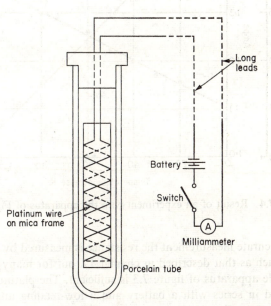

Figure 7.3 Circuit of a platinum resistance thermometer.

High temperatures can be measured by what is called a resistance thermometer. This is essentially a length of platinum wire wound on mica and placed inside a silica tube so that it can be immersed in, say, hot furnace gases. It is suitable for use at temperatures up to 1275 K. In the case of platinum and in view of the high possible value of $(t_2 - t_1)$ the above expression is not sufficiently accurate and it is necessary to introduce a second constant β. At any temperature we then have

$$R_2 = R_1 \left(1 + \alpha t + \beta t^2\right)$$

where α and β are given values appropriate to the intial temperature.

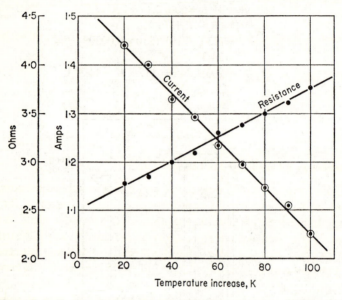

Figure 7.4 Result of an experiment with the apparatus of Figure 5.1.

For accurate measurement the resistance is measured by a 'bridge' circuit such as that described in chapter 19, but for many purposes the simple apparatus of figure 7.3 is sufficient. The platinum coil is connected in series with a battery and a low-reading ammeter or galvanometer. Changes in temperature cause changes in resistance and therefore in the instrument reading. For convenience, and again where extreme accuracy is not essential the instrument scale can be graduated in temperature degrees.

The apparatus of figure 5.1 can be used to determine the dependence of the resistance of a wire on its temperature. A coil of iron wire is suitable as its temperature coefficient is relatively high. A voltmeter, not shown, is connected to the terminals of the coil, the resistance at any moment being equal to the quotient of the p.d. at that moment and the current. If desired, the water can be heated by means of a bunsen burner turned low since readings can be taken more easily if the temperature rises slowly. The results of such an experiment are plotted in figure 7.4 and we see that over the range of temperature in the experiment the graph of resistance is a straight line, the equation

$$R_2 = R_1[1 + a(t_2 - t_1)]$$

therefore being accurate over this range. Thus from the graph we have

$$R_1 = 2.5, \quad R_2 = 3.8$$
$$\therefore \quad R_2/R_1 = 1.52$$
$$t_2 - t_1 = 100 \text{ K}$$
$$\therefore \quad a = \frac{1.52 - 1.00}{100} = 0.0052$$

Sodium as a Conductor for Electric Cables

Engineering is not only applied science, it is also applied economics. Britain is almost wholly dependent on foreign countries for its supply of copper. Sodium is the sixth most abundant metal in the earth's crust and can therefore easily be obtained in any industrially developed country. It also relieves any country without indigenous supplies of copper from dependence on world supplies for its current conducting materials.

Calling the market price per ton of copper 100, that of sodium is of the order 42. Again, the density of copper in g/cm^3 is 8.89 while that of sodium is only 0.97. Thus, in spite of the fact that the resistivity of sodium is 2.8 times that of copper, the cost of copper for a given current-carrying capacity is 7.6 times that of sodium.

The insulating material used for sodium cables is polyethylene, which has a melting point higher than that of sodium so that the molten sodium can be extruded directly into the insulating sheath. Since the sodium 'wets' the inner surface of the sheath there are no *voids* and therefore no danger of ionisation, which is one of the causes of breakdown in cables of orthodox construction.

A great advantage of the sodium-cored cable is its ability to withstand the very heavy short-duration currents produced by short-circuits. In a copper cable there is no change of state; that is, the copper does not melt, and consequently the whole of the joule heat is utilised in raising the temperature. A heavy short-circuit coinciding with a defect in the over-current protective device resulting in too long a time for the clearance of the fault can result in a temperature rise high enough to damage the insulation.

With a sodium-cored cable this danger does not exist because, owing to the low melting point of 97.8° C, and the use of an insulation which can withstand this temperature without damage, melting can be allowed; that is, there can be the change of state from solid to liquid. Now, during a change of state the temperature remains constant because the whole of the heat intake is utilised in producing this change. The most familiar case is that of water which remains at 100° C during the change of state from liquid to vapour. The heat necessary for the change of state is called the *latent heat*. In the case of sodium the joule heat will, if the temperature of 97.8° C is attained, be entirely utilised in supplying the latent heat of fusion and the temperature will remain at this value.

The mechanical properties of the sodium-cored cable are satisfactory, but they are not relevant to a book concerned only with electrical principles.

8 MAGNETISM

The appliances we call magnets are characterised by, among other properties which are considered later, several easily demonstrated properties:

(a) They possess at least two, sometimes more, regions which behave differently from the rest of the magnet.

(b) If dipped into light pieces of iron, such as iron filings, these adhere in large numbers to these special regions, but in very small numbers elsewhere.

(c) If an elongated magnet possessing these regions at its ends is suspended so that it can swing freely in a horizontal plane it will set itself in the magnetic meridian at the place where the experiment is made. In other words, it will set itself in a north-south direction.

(d) No matter how often it may be deflected from this chosen position, if it is left to itself it will take up this favoured position. Furthermore the end which pointed to the north will again point to the north no matter how often the experiment is repeated. Obviously the other end will point to the south. The north pointing end is called the north (N) pole, and the south pointing end the south (S) pole.

(e) If the N pole of a second magnet is presented, first to the N pole and then to the S pole of the suspended magnet, the N pole will be repelled and the S pole attracted. If the S pole of the second magnet is presented then, in this case, the N pole of the suspended magnet will be attracted, and the S pole repelled. It follows that like poles repel one another, and unlike poles attract one another.

(f) If a piece of iron which is not in the magnetic state is presented it will be found that there is attraction of both N and S poles.

(g) If pieces of material such as wood, glass, brass etc. are presented there will be no attraction at either end, showing that these materials are non-magnetic.

These simple experiments show that, from the magnetic point of view there are three kinds of material: magnets; materials which, although not themselves magnets, are attracted by magnets and thus possess magnetic properties; and non-magnetic materials.

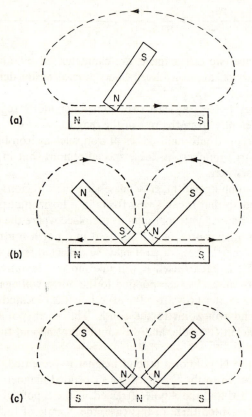

Figure 8.1 Magnetisation by stroking.

Making a magnet

If a magnet is to retain its special property permanently it must be made of the appropriate material; for example steel. If a permanent state of magnetisation is not necessary then the material must be such that it loses its magnetic property as soon as the magnetising agency is removed. Such a material is 'soft' iron. Magnets can be made:

(a) by stroking. This is a method of making a permanent magnet, and there are two variants as shown in figure 8.1. In figure 8.1a a magnet NS is moved along the steel bar AB to be magnetised so that the bar is stroked by one pole, the N pole in the figure. The magnet is then moved back to the starting point, its path being that of the dotted line. The process is repeated a number of times and the end A becomes a N pole and the end B a S pole. We see that the polarity at the end where the stroking magnet leaves is opposite to that of the stroking pole. Figure 8.1b shows a similar method using two, instead of only one, stroking magnets.

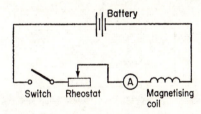

Figure 8.2   Circuit of a simple electromagnet.

If this second method is used with both stroking poles of the same polarity, say N, then the two ends of the magnet so made will be of the same polarity, S in this case (figure 8.1c). The middle will be of N polarity. This additional pole is called a consequent pole. Sometimes it is convenient to use consequent poles in the magnetic system of electrical machines. The stroking method does not produce powerful magnets.

(b) by electric current.

First consider temporary magnetisation.

A length of wrought-iron rod has wound on it a close coil of copper wire—18 gauge single cotton covered (s.c.c.) is suitable—and a short length is left clear at each end (figure 8.2). Connect the coil in series with a variable rheostat, ammeter and a switch, and use as current source a battery of two lead-acid accumulator cells in series. With the rheostat adjusted so that the whole of its resistance is in circuit close the switch. Then dip the ends of the rod into the iron filings and it will be found that many filings adhere. This shows that the iron bar is a magnet so long as the current is flowing. Now open the switch and it will be seen that most of the filings fall off, showing that with 'soft' iron magnetisation is dependent on the presence of the electric current in the coil.

This experiment demonstrates the fundamentally important fact that magnetism is not an isolated phenomenon but that it is intimately associated with electricity. This is discussed more fully in chapter 9.

(c) By 'Flashing'. The apparatus is similar to that of figure 8.2 except that the supply is taken from a d.c. mains supply and there-

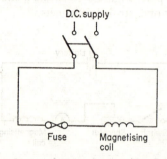

Figure 8.3   Circuit for making a permanent magnet by 'flashing'.

fore a double-pole switch is necessary in order to give complete isolation when the switch is open. Also a fuse is included and care must be taken to ensure that it is of such capacity that it will blow before the main fuse (figure 8.3). No controlling rheostat is used since the object is to secure a very rapid rise of current and then its sudden cessation, not a sustained current as in case (b). The specimens to be magnetised can be bars of hard steel and it will be found that, after flashing, they have become permanent magnets. If several such magnets are made they can be used for the mapping of magnetic fields, as described below. At first this experiment should be performed under supervision.

The above process is essentially the same as the industrial method of making permanent magnets. For industrial purposes there are two important requirements in permanent magnet steel. (i) The strength of a magnet shall be as great as possible for a given volume of material; this is because the special materials used for this purpose are very expensive. (ii) The magnet must maintain its strength with little loss, preferably none at all, this being of fundamental importance in the case of magnets used in electrical measuring instruments. The gradual diminution of ordinary steel magnets, called ageing, is due to the fact that there is a certain amount of carbon present and this causes the molecular structure of the material to change as time goes on. For important applications alloys containing cobalt, nickel, or other metals are now used, their carbon content being negligible.

Magnetic Fields

We defined the strength of an electric field at any point as the force on a probe charge at that point divided by the quantity of electricity on the probe. We defined an electric line of force at a point as a line to which the direction of the force acting on the probe charge is tangential.

In the case of a magnetic field we can not make an analogous definition because there is no such thing as an isolated magnetic pole. An electric charge is the result of an accumulation or deficit of electrons, and these electrons can be transferred from one body to another. Magnetisation is not the result of an accumulation, or loss, of magnetic material because there is no material involved. It is, as we shall see, a rearrangement of groups of molecular magnets, called domains, such that all the domains point in the same direction. A satisfactory definition of a magnetic field must therefore be left to chapter 9. In the meantime we can define the direction of a magnetic field at any point as the direction in which a very small freely suspended magnet will point. This indicates a simple method of plotting a magnetic field.

Compass needle method

The method of determining the field for a single bar magnet is illustrated in figure 8.4. A small compass needle is placed close to one pole and when it is quite steady the position of its ends are marked by dots. The needle is then moved a short distance away and adjusted until the end nearer the magnet is just at the second dot. The position of its other end is then marked by another dot. The process is repeated until the needle has reached the magnet again, or has reached the edge of the paper. The dots are then joined by a smooth

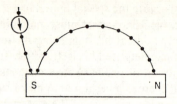

Figure 8.4   Field plotting by the compass method.

curve. Using a different starting point a second line is obtained, and so on until a complete map of the magnetic field right up to the boundary of the paper has been obtained.

Figure 8.5 shows the map of a field obtained in this way. It is of the combined effects of a bar magnet and the Earth's magnetic field. The N pole of the magnet pointed to the North. It will be seen that, close to the magnet the shapes of the lines of force are controlled almost entirely by the magnet alone, but towards the boundaries

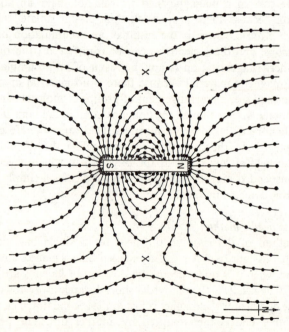

Figure 8.5   Plot of the magnetic field of a bar magnet pointing North, made by the point-by-point method.

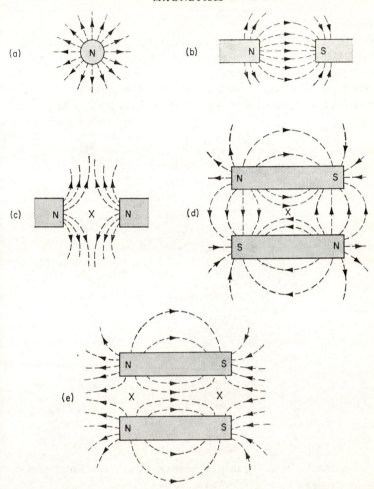

Figure 8.6 Various magnet arrangements suitable for field plotting. The single pole of (a) is obtained by mounting the magnet vertically. Note the null points at the places marked X.

where the magnet's field is weak they tend to become the parallel North–South lines due to the earth alone. At the two places marked X it is clear that the effects of the magnet and the earth just neutralise one another. Such points are called neutral points.

It is a good plan to make maps of as many fields as possible, and figure 8.6 suggests a number of suitable arrangements.

Iron-filing Method

A second method, which gives permanent records, is the iron-filing method. A sheet of paper impregnated with paraffin wax is placed on a sheet of card and this is placed over the magnet or magnets for which the shape of the field is required. The paper is then evenly sprinkled with iron filings, conveniently from a tin holder with a

Figure 8.7   Iron filing map of the magnetic field of a bar magnet.

uniformly perforated lid. If we could achieve a frictionless surface the filings would immediately set themselves along lines of force; as this is not possible the card must be tapped gently. This necessity of overcoming friction is a feature which limits the accuracy of the method, and it is therefore only suitable for fairly strong fields. When the filings give a satisfactory figure the surface of the paper is very gently warmed by means of a bunsen flame to melt the wax and secure the filings. Figure 8.7 shows an iron-filing map of the field of a single bar magnet obtained in this way.

Direction of magnetic lines of force

The direction of a line of force at any point is the direction of the magnetic force at that point on the poles of a very small magnet—we could call it a probe magnet—which is free to move. The sense of the magnetic force at a point is the direction of the force acting on the N pole of the probe magnet, and the opposite direction to the force acting on the S pole. Thus, outside the magnet the lines of force act from north pole to south pole. The theory of magnetism, discussed below, indicates that the lines of force also pass through the magnet itself, but this time from the south pole to the north pole as indicated in figure 8.8. Hence, unlike electric lines of force which

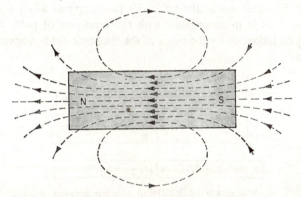

Figure 8.8 Magnetic lines of force pass through the magnet from the south to the north pole.

start at the surface of a positively charged conductor and end at the surface of a negatively charged conductor, not passing through the conductor at all under static conditions, magnetic lines of force are closed lines.

The north pole of a magnet is the region where the lines of force emerge and the south pole is the region where they enter. This is by far the best concept of a magnetic pole because it can be used in connection with magnets and electromagnets of any shape whatever.

Properties of magnetic lines of force

The forces between magnets can be explained by assuming that magnetic lines of force possess certain properties. Of course, lines of force, either electric or magnetic, do not exist, but to imagine that they do has proved so useful that the electrical engineer uses them as though they were real. Thus, when designing an electric generator or motor, it is decided at a very early stage how many lines of force each pole of the machine will provide, while the sizes of many of the parts are fixed by the number of lines of force to each unit area of their cross-section. The assumed properties are as follows:

(a) Lines of force tend to contract in length. They act, in this respect like stretched elastic threads. This corresponds to the experimental fact that if lines of force pass from one body to another, for example the N pole of one magnet and the S pole of another, the bodies attract one another.

(b) Neighbouring lines of the same directional sense repel one another. In other words the field tries to occupy as much room as possible. This is in accordance with the repulsion of poles of like polarity as indicated by the maps of the magnetic fields, for example
    ure 8.6.

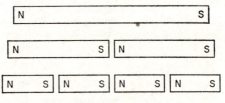

Figure 8.9   Subdivision of a bar magnet.

(c) As with electric lines of force there can be no intersection of two lines since this would mean that, at the point of intersection, the field acted in two different directions.

Theories of Magnetism

If a bar magnet is divided into two halves it will be found that each half is a normal bar magnet with N and S poles. At the break one half will have a N pole, the other a S pole. The same will happen if the halves are divided into quarters, and so on for as long as division is possible. A convenient magnet for this purpose is a length of straightened clock spring which has been magnetised by the stroking method or, preferably, by flashing. This is illustrated in figure 8.9.

If we imagine this process repeated until, eventually, the magnet is sub-divided into its component molecules, there seems good reason to assume that these molecules are themselves minute magnets. This is the *molecular theory of magnetism*. As stated in this way the theory is incomplete because it does not state how the molecules receive their magnetism. The explanation is given in chapter 9.

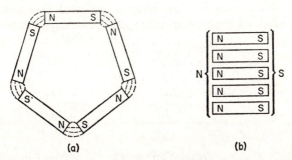

Figure 8.10 (a) Closed magnetic chain. (b) The magnets in the chain rearranged to form a composite bar magnet.

How then can a piece of magnetic material be unmagnetised if its molecules are all permanent magnets? Figure 8.10a shows five bar magnets arranged in what is called a closed magnetic chain. Adjacent poles are of opposite polarity and are close together. Hence magnetically, the magnets are joined by the lines of force joining adjacent poles. With the poles very close together the magnetic field of the whole group will be exceedingly weak except quite close to the poles. In other words, with respect to the space outside the group behaves as though it was unmagnetised. In addition, the force of attraction between adjacent poles will be very strong showing that such a group is very stable.

If the group can be broken up and the magnets all oriented in the same direction, as in figure 8.10b, then they will behave like a composite magnet and the field will extend for a considerable distance. If neighbouring groups can be opened, then the magnets

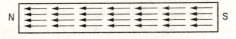

Figure 8.11 Open chains of molecular magnets, all having the same orientation.

can arrange themselves in a series of open magnetic chains, as in figure 8.11. The whole group will now have N polarity at one end and S polarity at the other, and it will thus behave like a bar magnet.

This is a model of what takes place when a bar of unmagnetised material is converted into a magnet, the difference being that the separate magnets of figure 8.10 are to be replaced by the molecular magnets of the material. According to accepted theory it is not the individual molecular magnets which form chains but groups of

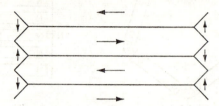

Figure 8.12   Magnetic domains arranged in closed chains in unmagnetised material (partly diagrammatic).

molecules called domains. Figure 8.12 shows, in simple form, a possible arrangement of domains when the material is unmagnetised. The chains are closed by the small domains at the ends which are called closure domains. The process of magnetisation is essentially as described for groups of magnets, the essential difference being that, during magnetisation some of the domains can increase at the expense of their neighbours, showing that the domain boundaries are not rigidly fixed.

Ferrites

When a ferromagnetic material is subjected to alternating current magnetisation, a loss of power takes place because of currents, called eddy-currents, which are induced in the material by the changing magnetic flux. This loss can be reduced by laminating the material; that is, building it up in the form of thin strips (chapter 16). If there are n strips then, for the same magnetisation, the loss is $1/n^2$ of that in solid material. Also the loss is directly proportional to the frequency of reversal.

With the sheet material used in electrical machine manufacture the loss can be kept to an acceptable value so long as the frequency is low. For radio transmission frequencies can be reckoned in megacycles, and in such cases ordinary ferromagnetic materials, however thin, experience exceedingly high losses. A class of ferro-magnetic material, called ferrites, has been developed which can be operated satisfactorily at such high frequencies. The chemical composition is of the form MO. Fe_2O_3, in which M can be any one of a number of divalent ions such as copper, zinc, cadmium, magnesium, or even iron itself. Mixed ferrites are possible, for example $(\frac{1}{2}$ Cu $\frac{1}{2}$ Zn) $O.Fe_2O_3$, so that a whole range of ferrites with a wide range of physical properties is possible. Thus ferrites are not metals but compounds of which one constituent is iron. Ferrites are made by heating the appropriate mixture for several hours at temperatures of the order of 1300° C, the actual value depending on the mixture itself.

As used for the magnetic cores of very high-frequency apparatus the material is in powder form, giving a subdivision very much greater than the lamination of solid material. As a result, the e.m.f. induced in each eddy-current path is minute, and this, combined with the very high resistivity of up to 10^6 Ω m, limit the magnitudes of the eddy-currents and the eddy-current loss. Because of their high resistivity ferrites belong to the class of intrinsic semiconductors (chapter 18).

9 THE MAGNETIC FIELDS PRODUCED BY ELECTRIC CURRENTS

The fact that an electric current always produces a magnetic field, no matter what the shape of the circuit may be, was discovered accidentally by Oersted in 1820. He was demonstrating what he believed to be the non-connection between electricity and magnetism. On arranging a current-carrying wire parallel to a compass needle he was surprised to find that the needle was deflected. On reversing the current the defection was reversed.

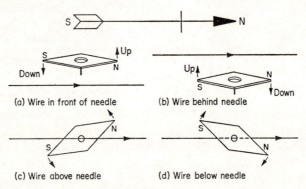

(a) Wire in front of needle

(b) Wire behind needle

(c) Wire above needle

(d) Wire below needle

Figure 9.1 Experiments to demonstrate the magnetic effect of a straight conductor when carrying current.

It is instructive to repeat Oersted's experiment in the following way. Connect a length of copper wire, say about one metre of no. 18 s.c.c., to a Leclanché cell, and include a switch so that current need not flow when observations are not being made. Arrange a length of about 20 cm of the wire horizontally near to a compass needle, as in figure 9.1. With current flowing hold the wire in the following positions: (a) at the same level as the needle and in front of it, (b) behind the needle, (c) above the needle, (d) below the needle. In cases (a) and (b) one end of the needle will dip downwards and the other end rise upwards, although the movement may be very slight with a weak current. In cases (c) and (d) the needle will be deflected from its north–south direction.

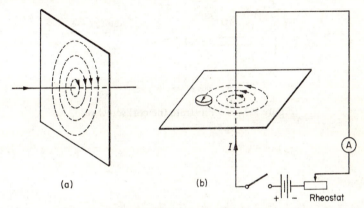

Figure 9.2   Magnetic field due to a straight conductor.

The deductions from the above experiments are that the lines of force of the field set up by the current are horizontal above and below the wire and at right-angles to it, and that the lines of force are vertical at the sides of the wire. Owing to radial symmetry these conditions can only be fulfilled if the lines of force are circles having their centre in the wire, and such that the wire is at right-angles to the plane containing them. This is illustrated in figure 9.2 which is drawn for one such plane only. If the wire is passed vertically through a horizontal card, as in figure 9.2b the field can be plotted both by the compass needle and by the iron-filing method. In these cases, to be successful, the current should be several amperes, the source therefore being a battery of secondary cells. The compass needle method also shows that when the current is reversed the field is reversed.

Direction of the field (and magnitude)

The simplest rule is the corkscrew rule. The above experiments show that when the conductor is straight the lines of force are circular. In the corkscrew we have a combination of straight-line and circular motion. The rotation of the handle imparts to the whole appliance a straight-line motion whose direction is dependent on the direction of rotation. It so happens that with a corkscrew or any right-handed screw the two motions correspond to the straight line current

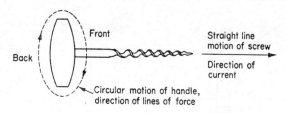

Figure 9.3   To illustrate the corkscrew rule.

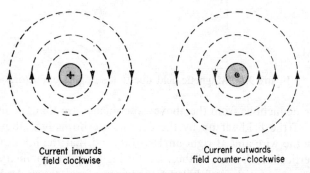

Figure 9.4   Directions of the lines of force dependent on the direction of the current.

direction and circular magnetic field direction. Let the bodily motion of the corkscrew represent the direction of the current in a straight conductor; then the direction in which the handle must be rotated gives the direction of the lines of force.

For the purpose of a diagram it is often necessary to show the current in cross-section; the current will then be inwards or outwards and the lines of force will be circles in the plane of the paper (figure 9.4). The rule shows that for inward currents the lines of force are clockwise, while for outward currents they are counter-clockwise.

The magnitude of the field strength is proportional to the current and inversely proportional to the distance r from the centre

$$\therefore \ B \propto I/r$$

The constant is equal to 2×10^{-7}

$$\therefore \ B = 2 \times 10^{-7} \ I/r$$

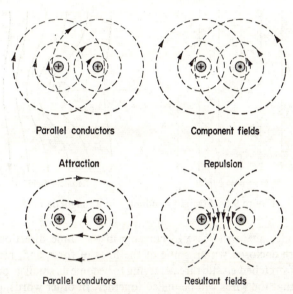

Parallel conductors Component fields

Attraction Repulsion

Parallel condutors Resultant fields

Figure 9.5 Magnetic fields due to parallel conductors.

Field due to two parallel conductors

Each conductor will, if acting alone, set up a field consisting of circular lines of force. If there are two conductors then there will be two components, and the field will be the resultant of two sets of intersecting circles. There are two cases depending on whether the two currents are in the same direction or in opposite directions. Figure 9.5 illustrates both cases, the top figures showing the component fields and the bottom figures the resultant fields. With equal currents in the same direction the field at the point midway between the wires is zero, this therefore being a null point. At all points between the wires the component fields oppose one another so that in this region the field is weak. Outside the wires the component fields act together and at large distances the effect is very approximately that of a single conductor.

With currents in opposite directions the component fields augment one another between the wires so that, in this region the field is strong. Outside the wires the component fields are in opposition, and at large distances the field is exceedingly small.

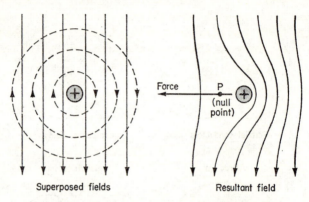

Figure 9.6   Straight conductor in a magnetic field.

Consider again the case of currents in the same direction. Both wires are enclosed within some of the lines of force and, regarding these as stretched elastic threads trying to become as small as possible, we see that the wires will be pulled together. In other words, parallel conductors carrying currents in the same direction attract one another. In the case of currents in opposite directions there are no lines which enclose both conductors but, in between the two, there is a strong field consisting of lines of force all of the same sense. Such lines of force repel one another laterally and, in so doing, they set up a force of repulsion between the wires. In other words, parallel conductors carrying currents in opposite directions repel one another. As we have seen, the force between parallel current-carrying conductors is the basis of the SI definition of the ampere.

Force on a single conductor in a magnetic field

This is another example of the superposition of two component magnetic fields. Figure 9.6 shows in cross-section a conductor with an inward flowing current. The lines of force produced by this current are the circles shown dotted. The conductor is at right-angles to the lines of force of a field, the main field, produced by another agency, say an electromagnet, and the field of this is a uniform distribution of parallel lines of force having a downward sense. Combining the two fields we see that on the right-hand side of the conductor the main field is strengthened whereas on the left-hand side it is weakened. In a sense, some of the lines of the main field are displaced from the left-hand to the right-hand side of the conductor resulting in the distortion shown in the right-hand figure. Since lines of force are always in a state of tension the bent lines of force are

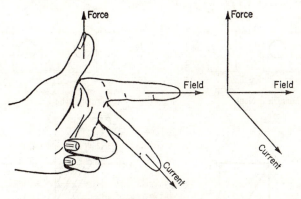

Figure 9.7   The left-hand rule.

rather like the stretched elastic of a catapult and we see that in this case there is a mechanical force on the conductor which acts at right-angles to the direction of the main field, and also at right-angles to the conductor itself.

Thus we have three directions, each at right-angles to the plane containing the two wires, the OX, OY and OZ co-ordinate axes. The relationship between these three is given by the left-hand rule, figure 9.7. To apply the rule arrange the first and second fingers and the thumb of the left hand all mutually at right-angles. Point the first finger in the direction of the main field, and the second finger in the direction of the current; then the thumb will point in the direction of the force. In applying this rule to the case of figure 9.6 hold the book so that the page is in a vertical plane. Point the first finger downwards so as to correspond with the direction of the main field, point the second finger towards the paper so as to correspond with the direction of the current. The thumb will then point to the left, which is the direction of the force.

It is frequently stated that the bending of the lines of force round a current-carrying conductor is the cause of the force which acts on the conductors of any electric machine of normal construction. This is not correct because the conductors are all housed in slots in which there is very little magnetic flux (by flux we mean the total number of

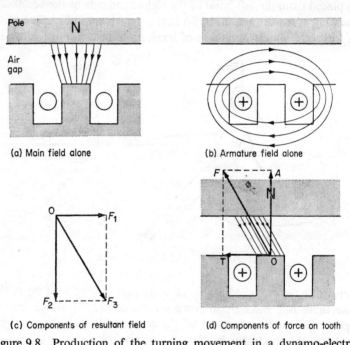

(a) Main field alone

(b) Armature field alone

(c) Components of resultant field

(d) Components of force on tooth

Figure 9.8   Production of the turning movement in a dynamo-electric machine.

lines of force). Hence there are, at most, very few lines of force to bend, practically the whole of the field passing through the teeth. Figure 9.8a shows two slots with one tooth in between and one conductor per slot. With no current in the conductors, only one per slot being shown for simplicity, the flux from the pole crosses the air gap in tufts which are, on average, radial. Figure 9.8b shows the flux which the conductors would produce if there were no main flux. The lines of force pass up the tooth on the left, cross the air-gap, down the tooth on the right and then across the iron below the teeth.

We see that the conductor alone produces in the air-gap a field which acts from left to right. It is represented by the force OF_1 in figure 9.8c. The main field of the flux entering the tooth acts downwards, OF_2 in figure 9.8c. The resultant of these two is the field OF_3. Consequently the lines of force which enter the tooth will take the general direction of OF_3, the tufts of flux thereby becoming inclined as shown in figure 9.8d. The length of the path of the lines of force is now greater than it was in figure 9.8a and consequently there is an attempt to shorten the path, again because of the mechanical property of the lines of force. For the same reason the force OF acting on the top of the tooth is inclined, its direction being parallel to OF_3. This force has two components; OA a radial force which causes the pole to attract the armature but makes no contribution to the output of the machine, and OT the tangential component. It is the sum of all these tangential components multiplied by the radius which gives the turning moment of the machine. It takes place both when the machine is acting as a generator and when it is acting as a motor. In short, the turning moment is due to forces on the teeth, not forces on the conductors.

In order to avoid unnecessary complication it is often necessary to draw a diagram with the conductors in air, not in slots, but it should be remembered that this is merely a convenience.

Field of a single circular turn

If we draw a section of such a coil in a plane through the axis we merely have two circles representing conductor sections carrying currents in opposite directions. As far as this plane is concerned the magnetic field is similar to that of the bottom right-hand figure of figure 9.5. In the plane of the coil the lines of force are all parallel to the axis and therefore perpendicular to this plane. The lines diverge very rapidly as the distance from this plane is increased. Figure 9.9a. If we look at the coil from one side the current circulation will be clockwise, from the other side it will be counterclockwise. The face where the lines come towards the observer can be regarded as the N face; the face where the lines recede from the observer can be regarded as the S face. The coil can, in fact, be regarded as a flat magnet of exceeding small axial length. Figure 9.9b shows a very simple aid to memory.

Field of a solenoid

The word *solenoid* is used to denote a coil whose axial length is much greater than the diameter of the turns of which it is composed. In a sense it is a number of single turns in series. The portion of the field in the plane of a single turn consists of lines of force parallel to the axis. Hence with many turns the lines of force inside the solenoid are lines parallel to the axis, like the lines of force inside a bar magnet. Outside the solenoid they take paths also like those due to a bar

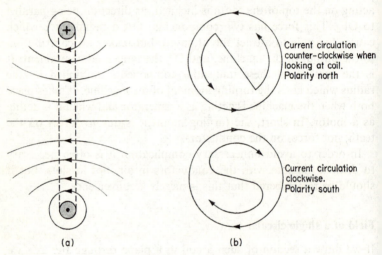

Current circulation counter-clockwise when looking at coil. Polarity north

Current circulation clockwise. Polarity south

(a)   (b)

Figure 9.9   Field due to a single turn.

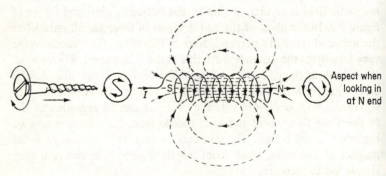

Aspect when looking in at N end

Figure 9.10   Magnetic field of a solenoid.

magnet. In fact, a solenoid carrying current behaves exactly like a bar magnet, the end where the lines of force emerge being the N pole and the end where they enter, the S pole (figure 9.10).

Here again we have a combination of straight line and circular phenomena, but this time it is the current which flows in circles while the lines of force inside the solenoid are straight. The corkscrew rule applies to this case also, the circular motion of the handle corresponding to the current circulation, and the bodily motion to

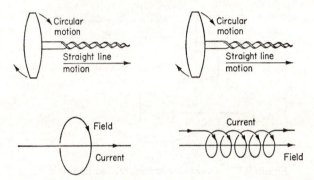

Figure 9.11 Application of the corkscrew rule to both a straight conductor and a solenoid.

the direction of the lines of force inside the solenoid. The application of the rule to both straight and circular conductors is illustrated in figure 9.11.

Magnitude of the Force

The production of a mechanical force on a current-carrying conductor placed in a magnetic field is the fundamental operating principle of the electric motor.

If B = flux density in si units
  I = conductor current in amperes
  l = conductor length in m (at right-angles to the field)
  F = force on the conductor in newtons, then
  $F = BIl$

This relationship is the basis of the definition of the unit of magnetic flux density, which is:

The unit of magnetic flux density is such that a long conductor carrying a current of one ampere, placed at right-angles to a field of unit density experiences a force of one newton per metre length.† The name of the unit is the tesla (symbol T). We shall see that it is also useful to think of flux density as so many magnetic lines of force crossing unit area of normal cross-section.

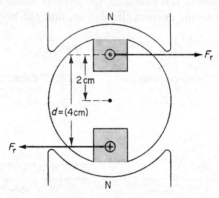

Figure 9.12   Generator armature. See Example 9.3.

Example 9.1.   A conductor 20 cm long, carrying 100 A, is placed at right-angles to a magnetic field of 5×10^{-3} T. Calculate the force acting on it.

$$T = 5 \times 10^{-3}; \; I = 100 \text{ A}; \; l = 20 \times 10^{-2}$$
$$\therefore \; F = 5 \times 10^{-3} \times 10^2 \times 20 \times 10^{-2} = 10^{-1} \text{ N}$$

Example 9.2.   In a motor the field strength is 0.6 T. The active length of each conductor is 40 cm and each carries 50 A. Calculate the force on each conductor

$$T = 0.6; \; I = 50 \text{ A}; \; l = 40 \times 10^2 = 0.4 \text{ m}$$
$$\therefore \; F = 0.6 \times 50 \times 0.4 = 12 \text{ N}$$

Example 9.3.   The armature of a two-pole generator driven by hand through gearing is of the so-called H section, that is, there are two large slots housing a single coil of many turns (figure 9.12). The centre of each slot is 2 cm from the axis, each slot carries 200 conductors and the field strength is 0.3 T. The length of the active

† The SI definition is given on p. 96.

conductor is 5 cm. If the current is 0.05 A, what is the torque acting as the armature?

$F = BIl$ per conductor
$B = 0.3 = 3 \times 10^{-1}$ T; $I = 5 \times 10^{-2}$ A; $l = 5 \times 10^{-2}$ m

$\therefore \quad F = 3 \times 10^{-1} \times 5 \times 10^{-2} \times 5 \times 10^{-2} = 7.5 \times 10^{-4}$ N

$\therefore$ Force per slot
$F_s = 7.5 \times 10^{-4} \times 200 = 0.15$ N

Diameter $d = 4 \times 10^{-2}$ m

$\therefore$ Torque $= F_s d = 0.15 \times 4 \times 10^{-2} = 6 \times 10^{-3}$ N m

Two Parallel Conductors

Let the distance between centres be r metres and let the currents be I_1 and I_2 amperes. The conductor carrying I_1 amperes will produce at a distance r from its centre a magnetic field of strength

$$B = 2 \times 10^{-7} I_1/r \text{ tesla}$$

where (2×10^{-7}) is the value of the numerical constant which relates the three entities when they are all expressed in SI units. But the second conductor is in this particular field, and therefore the force acting on it

$$F = (2 \times 10^{-7} I_1/r) I_2 l.$$

and obviously this is also the force on the conductor carrying current I_1. If the two currents are equal then

$$F = 2 \times 10^{-7} I^2 l/r \text{ newtons}$$

It will be realised that this is the basis of the definition of the SI unit of current, stated on page 33.

Example 9.4.   Two parallel wires, in air, each 10 m long and 0.1 m apart (centre to centre) carry a current of 100 A. Calculate the force acting on each wire.

With the ratio of length to spacing being so great we can neglect the effects of the ends (the definition of the ampere refers to infinitely long wires)

$$F = 2 \times 10^{-7} \times (100)^2 \times 10/0.1 = 0.2 \text{ N}$$

10 THE MAGNETIC CIRCUIT

The path of an electric current is called the electric circuit. Assuming a complete circuit the Ohm's law relationship is

$$\text{current } (I) = \frac{\text{e.m.f. } (E)}{\text{resistance } (R)}$$

The path of the magnetisation of any appliance is called the magnetic circuit, and there is a similar relationship to that of Ohm's law in the electric circuit. It is

$$\text{flux } \Phi = \frac{\text{m.m.f. } (F)}{\text{reluctance } (S)}$$

The word flux is a misnomer because in the magnetic circuit, there is no flow of particles of magnetism. We must therefore be content with the SI definition, although it is based on phenomena which have not yet been described. The unit is called the weber, defined as follows.

The weber is the flux which, linking a circuit of one turn, produces in it an electromotive force of one volt as it is reduced to zero at a uniform rate in one second.

The magnetomotive force is produced by the current in the magnetising coil. Consider a flat coil of one turn (figure 10.1a). If it carries a current I it will produce a flux Φ through a given area. If the current is doubled, figure 10.1b, the flux in the same magnetic path will be 2Φ. If the coil has two turns, figure 10.1c, and the current is again I, the flux will be 2Φ, as in figure 10.1b. If the coil has two turns and carries a current $2I$, figure 10.1d, then the flux in the same path will be 4Φ. Thus the flux is proportional to the product of N and I, namely NI. It is called the ampere turns. The SI system of

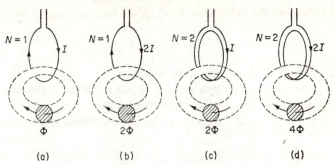

Figure 10.1 The effect of the magnitude of the current and of the number of turns on the m.m.f. of a coil.

units is so designed that the m.m.f. is equal to the ampere turns. The symbol for the entity is F. Since a number, as the number of turns, has no dimensions, the unit of m.m.f. is the ampere A. Again, the e.m.f. produces throughout the circuit an electric field of strength which we will now denote by $\mathcal{E}$, where, for a uniform circuit of length l

$$\mathcal{E} = E/l$$

Analogous to the electric field strength $\mathcal{E}$, we have the magnetic field strength or magnetising force, H. We therefore have

$$H = F/l$$

the unit being the ampere per metre, A/m.

The third component is the magnetic reluctance. For the electric resistance we have

$$R = \rho l/a$$

If we put $\rho = 1/\sigma$ where σ is the conductivity then

$$R = l/\sigma a$$

The entity analogous to electrical conductivity is *magnetic* permeability, symbol μ, giving

$$S = l/\mu a$$

The permeability is a composite quantity, its nature being explained on p. 98. We can say that it is a constant for non-magnetic materials, but not for ferromagnetic materials such as iron and steel.

There remains the entity flux density, symbol B. It is the quotient of the flux and the area through which the lines of force of the flux pass

$$\therefore \ B = \Phi/a$$

The SI definition of magnetic flux density is:

The unit of magnetic flux density, called the tesla, is the density of one weber of magnetic flux per square metre.

The following example gives an idea of the magnitudes involved.

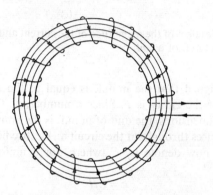

Figure 10.2   Ring-shaped specimen with toroidal magnetising coil.

Example 10.1.   A 300 kW generator with 8 poles has a flux of 0.073 Wb. The area of the pole face is 0.11 m². What is the mean value of the flux density over the pole face?

$$B = \Phi/a = 0.073/0.11 = 0.67 \text{ T}$$

For a two-pole modern turbo alternator the flux per pole will be of the order of 2 Wb. We see that the weber is an exceedingly large unit.

Magnetisation

As we have seen, a bar of suitable material can be magnetised by placing it inside a solenoid. The degree of magnetisation will depend on the magnitude of the current. A bar has ends, and therefore the flux leaving at the N end has to take an air path before re-entering at the S end and some of the ampere turns are required for this air path. This difficulty is overcome by using a ring-shaped specimen on which is wound a ring-shaped coil or toroid, figure 10.2. If l is the mean circumference then, for any value of I

$$H = AT/l = NI/l$$

The ferromagnetic material can be regarded as an assembly of domains; each domain is a minute magnet, the unmagnetised state of the specimen depending on the grouping of the domains in closed chains. The stages in the rearrangement of a chain during progressive increase in H are shown diagrammatically in figure 10.3. Figure 10.3a shows the chain completely closed. If a magnetising force H is applied, a small value of H will produce a small deformation as

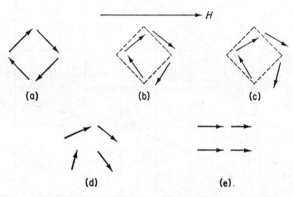

Figure 10.3   Stages in the conversion of closed magnetic chains to open chains.

in figure 10.3b. This will result in weak magnetisation. If it is doubled the deformation will be roughly doubled so that the magnetisation, as given by the value of B, will be about doubled. Thus at first the flux density B is sensibly proportional to H, and therefore to I, so that, initially, the curve will be almost a straight line through the origin (as shown in figure 10.4). A further increase in H causes a break-up of the chain as in figure 10.3c after which rotation of the domains into the direction of H will be much easier. There is, therefore, a rapid increase in B. After this, further alignment along the direction of H can have little effect so that, finally, the curves rise very slowly. With complete alignment the magnetisation of the specimen induced by application of H can increase no further and the specimen is then saturated. Since the magnetising force H makes its own contribution to the total flux the value of B continues to rise, but only slowly.

Curves for a few important magnetic materials are given in figure 10.4, from which it will be seen that with these materials the initial

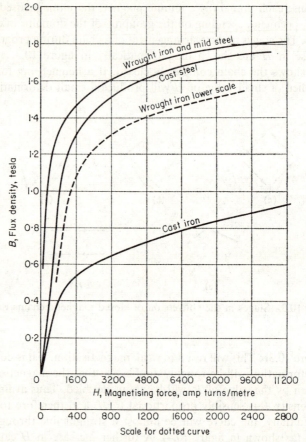

Figure 10.4   Magnetisation curves for a few magnetic materials.

portion is very small, the chains being broken at very small values of H. For a given value of H it will be seen that the flux density for cast iron is only about one-half of that for the other materials. Cast iron is therefore rarely used in electrical machines for those portions which have to carry the magnetic flux.

Magnetic reversals. Hysteresis

If a specimen is magnetised by a gradual increase in H its successive states will be indicated by a curve such as OAB, figure 10.5. Since resistance is encountered in breaking the magnetic chains, work will be done. If H is now reduced to zero, and if the material retraced the curve in the opposite direction (BAO), then this work would be returned to the source. Actually the curve during the change to $H=O$ is more like curve BC. The specimen is still magnetised by the amount OC, which is called the *retentivity*. To demagnetise completely the

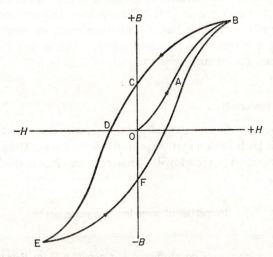

Figure 10.5   A cycle of magnetisation.

specimen a reversed magnetising force equal to OD has to be applied, this being accomplished by reversing the current. The value OD is called the coercive force or coercivity.

Increasing the reversed H to the previous maximum brings the specimen to point E. Finally H is brought to zero (curve EF), then reversed again and brought up to the previous maximum value (curve FE), a closed cycle having been traced out. It can be shown that to put the material through this cycle of changes the source has to supply energy. The phenomenon is called hysteresis and the complete loop is called the 'hysteresis loop'. The shape of this loop shows three very important properties of the material:

(a) Retentivity. From the practical point of view this property is almost as important as magnetisation itself, for, without it, many modern appliances, such as d.c. generators, would be unworkable.

(b) Coercivity. Although a specimen may have retained magnetisation, as OC, when H is brought to zero, the criterion for its retaining magnetisation is the value of the coercivity. So for the manufacture of permanent magnets one essential property of the material is a large coercive force, OD. We have seen that the material must not lose its magnetism through ageing.

(c) Hysteresis loss. The armature cores of generators and motors and the cores of transformers are subjected to reversals of magnetism with a loss of energy called the hysteresis loss. This loss reduces the efficiency of the machine but, equally important, the work done in supplying it is converted into heat energy which contributes to the temperature rise in such systems.

Low loss materials

Low loss materials are essentially materials with a very low coercive force; this leads to thin hysteresis loops of small area. The properties and composition of a few low-loss materials are given in the following table.

Properties of some low-loss materials

Material	Composition	Coercivity, A/m	Retentivity, T
Pure iron	Fe	4.0	1.36
Permalloy (78.5 per cent nickel)	Ni, Fe, Mn	4.0	6.6
Mumetal (74 per cent nickel)	Ni, Fe, Cu, Mn	4.0	0.5
4.5 per cent silicon steel	Fe, Si	40.0	0.5
1 per cent silicon steel	Fe, Si	56.0	0.85
Permendur	Fe, Co	160.0	1.4
Annealed cast iron (for comparison)	Fe	880.0	0.55

Many of these materials have been developed for special purposes, since their cost makes them prohibitive for use in ordinary machinery. In transformers silicon steels are often used because the losses in the core are a much greater proportion of the total loss than in the case of rotating electrical machines. The high costs are indicated by the following: if we call the cost of annealed cast iron 1 then the costs of the others are permalloy—28, mumetal—27, permendur—69, silicon steel (4.5 per cent)—8 and (1 per cent)—6, sufficiently low to justify their use in transformer construction.

Permanent magnet materials

For these materials it is the demagnetising portion of the characteristic, the triangle OCD in figure 10.5, which is of importance since the major requirement is as large an area in this region as possible. It is also essential that the coercivity should be as high as possible. The following table gives the properties of a number of permanent magnet materials and, like some low loss materials, they are very expensive.

Properties of some permanent magnet alloys

Material	Composition	Retentivity, T	Coercivity, A/m
Alnico V cast	Fe, Co, Ni, Al, Cu	1.25	44 000
Alnico II cast	Fe, Co, Ni, Al, Cu (Cu various)	0.72	43 200
Alnico I cast	Fe, Co, Ni, Al	0.73	34 400
35 per cent cobalt steel	Fe, Co, C	0.95	20 800
6 per cent tungsten steel	Fe, W, C	1.00	6 400
3 per cent chrome steel	Fe, Cr, Mn, C	0.97	5 200

Magnetic Spin

We have seen that electrons are particles of negative electricity; they behave as though spinning on an axis, much as the earth spins on its axis. A spinning negative charge is an electric current since each point in it is travelling in a circle. If, on looking down on to the spinning electron its rotation is counterclockwise, then, since its charge is negative, it is equivalent to a conventional current in a clockwise direction (figure 10.6a). The spinning electron can therefore be represented by a minute circular current as in figure 10.6b. We see that it produces a magnetic field and, since the spin is permanent so is its field. The spins of all the electrons of an element are not all in

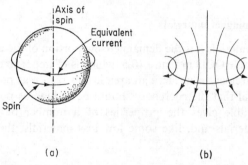

Figure 10.6 (a) Spinning electron. (b) Equivalent circular current.

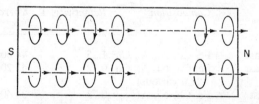

Figure 10.7 Arrangement of magnetic dipoles in a magnetised iron bar.

the same direction, and if there are as many clockwise spins as counterclockwise then the atom, as a whole, is not magnetic. In the case of the iron atom there are more spins in one direction than in the other and it is this which gives iron its special magnetic property.

Such a minute current circulation is called a magnetic dipole and we can therefore picture a bar magnet as consisting of open chains of dipoles, as in figure 10.7; not single atomic dipoles but groups (domains) of dipoles all having the same orientation.

Electromagnetic devices

Practical applications of electromagnetism are very numerous. If, through some calamity, magnetism or electric current no longer existed, civilisation as we know it today would come to an end. Some well known examples are detailed.

(a) Bells. The arrangement of two types of electric bell are shown in figure 10.8. The essential features are horse-shoe electromagnet M having an exciting coil on each limb; an iron armature A which is attracted when there is current in the coils; a contact 'make and break' C, and a spring S which causes an oscillating motion of the armature. In the continuous ringing type the closing of the circuit energises the magnet and the armature is violently attracted,

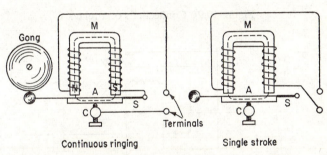

Figure 10.8   Connections scheme for electric bells.

striking the bell. At the same time contact is broken at C, the magnet loses most of its magnetism and the spring pulls the armature back, making contact again at C. The cycle is then repeated.

For some applications such as signalling, where a code consisting of a pre-arranged number of rings is in use, the trembling bell is not satisfactory and the single-stroke bell should be used. The construction is identical except that the screw C, used for adjusting the position of the armature relative to the magnet poles, is not used as a contact-breaker, and therefore the circuit can never be interrupted in the bell itself; when the circuit is closed only one ring is given.

It will be clear that the materials of the magnet core and armature must have the minimum retentivity to ensure that the armature will be released when required. To ensure that the armature does not stick, the screw C should be adjusted so that, in the attracted position, it does not quite touch the poles.

(b) Relay. This is another instrument employing an electro-magnet; it is a device for closing the circuit of some other electrical appliance. Figure 10.9 shows one type. It consists of a small electro-magnet M with an attracted armature A of bell-crank shape,

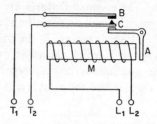

Figure 10.9   Connections of a relay.

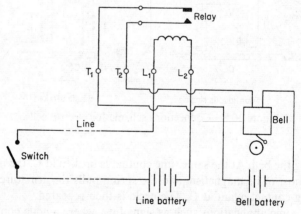

Figure 10.10   Application of relay to a bell circuit.

pivoted and spring controlled as in the single-stroke electric bell. In the off position the armature is pressed against an adjustable stop, but when it is attracted by the action of current through the magnet coils, led in at terminals $L_1 L_2$, the contacts B and C come together, thereby closing any circuit connected to terminals T_1 and T_2.

The relay is a very sensitive instrument and will operate with an extremely small current. Figure 10.10 shows the application to a bell circuit where the bell is such a long distance from the operating point that the resistance of the line wires prevents the current reaching the value required to operate the bell.

Example 10.1.   An electric signalling bell has a resistance of 20 Ω and requires a current of 0.5 A. The battery used has an e.m.f. of 25 V and resistance 2 Ω. If the wire used for the bell circuit has a restance of 46.5 Ω per kilometre of single run, what is the maximum operating distance possible without using a relay?

If l=distance in km, then

$$\text{Circuit resistance} = 2\ l \times 46.5 = 93\ l\ \Omega$$
$$\therefore\ \text{Total resistance } R_t = 93\ l + 2 + 20 = (93\ l + 22)\ \Omega$$
$$\therefore\ I = \frac{E}{R} = \frac{25}{93\ l + 22}$$
$$\therefore\ 0.5 = \frac{25}{93\ l + 22}$$
$$\text{giving } l = 0.3 \text{ km}$$

Magnetic calculations

If a certain coil carrying a stated current produces a flux Φ_1 when the former on which it is wound is unmagnetised and a flux Φ_2 when it is magnetised, the ratio Φ_2/Φ_1 is called the relative permeability, μ_r, of the magnetic material for the particular conditions of the test.

The electrical resistance of a uniform conductor is

$$R = \rho l/a = l/\sigma a, \text{ where } \sigma = 1/\rho \text{ is the conductivity}$$

The reluctance of a uniform magnetic circuit is

$$S = l/\mu a, \text{ where } \mu = \mu_0\ \mu_r$$

the product of the magnetic space constant μ_0 and the relative permeability μ_r. The magnitude of μ_0 is $4\pi \times 10^7$, this value having been chosen to suit the si system of units.

Example 10.2.   A ring of non-magnetic material of mean circumference 100 cm and cross-section 10 cm^2 has a uniform winding of 2500 turns. Calculate the flux produced by a current of 10A.

$$F = NI = 2500 \times 10 = 25\ 000 \text{ A}$$
$$l = 100 \times 10^{-2} = 1 \text{ m, and } a = 10 \times 10^{-4} = 10^{-3} \text{ m}^2$$
$$\mu = \mu_0 \text{ since } \mu_r = 1$$
$$\therefore\ \Phi = 25\ 000 \times 4\pi \times 10^{-7} \times 10^{-3}/1 = 3.14 \times 10^{-5} \text{ Wb}$$

Example 10.3.   An iron ring, 100 cm mean circumference has a cross-section of 10 cm². Its relative permeability is 500. If the magnetising coil has 500 turns what current will be required to produce a flux of 10^{-3} Wb?

$$l = 100 \text{ cm } = 1 \text{ m, } a = 10 \text{ cm}^2 = 10^{-3} \text{ m}^2$$
$$S = l/(\mu_0\mu_r\ a) = 1/(4\pi \times 10^{-7} \times 500 \times 10^{-3})$$
$$= 1.592 \times 10^6 \text{ AT/Wb}$$
$$\therefore\ NI = \Phi\ S$$
$$= 10^{-3} \times 1.592 \times 10^6 = 1.592 \times 10^3$$
$$\therefore\ I = 1.592 \times 10^3/500 = 3.18 \text{ A}$$

Magnetic circuit with air-gap

In order that a magnetic flux may be utilised it is necessary that there shall be a break in the iron circuit—the air gap—so that the appliance which utilises the flux may be inserted. Clearly the gap is in series with the iron if it is traversed by the same flux and consequently the total reluctance is the sum of the reluctance of the iron and of the air

$$\therefore\ S = S_i + S_a$$

the suffixes i and a referring to iron and air respectively. The relative permeability of air is unity

$$\therefore\ \mu_i = \mu_0\mu_r \text{ and } \mu_a = \mu_0$$
$$\therefore\ S_i = l_i/\mu_0\mu_r a_i \text{ and } S_a = l_a/\mu_0 a_a$$

Example 10.4.   Calculate the ampere-turns necessary to produce a flux density of 1 T in a magnetic circuit consisting of an iron path 50 cm long and an air-gap of 3 mm. The relative permeability of the iron at this value of T is 900.

The calculation is conveniently made in tabular form:

	Iron	Air
B	1.0 T	1.0 T
μ_r	900	1.0
$\mu_0\mu_r$	$4\pi \times 10^{-7} \times 900$	$4\pi \times 10^{-7}$
	$= 1.13 \times 10^{-3}$	
$H = B/\mu_0\mu_r$	$= 1/(1.13 \times 10^{-3})$	$1/(4\pi \times 10^{-7})$
	$= 8.85 \times 10^2$	$= 7.96 \times 10^5$
l	50 cm $= 0.5$ m	3 mm $= 3 \times 10^{-3}$ m
$F = Hl$	$8.85 \times 10^2 \times 0.5$	$7.96 \times 10^5 \times 3 \times 10^{-3}$
	$= 443$	$= 2388$

Total amp. turns $= 2831$

In the calculation it is assumed that the whole of the flux in the iron crosses the air-gap. In practice this is not the case because of leakage lines of force which take paths outside the gap.

Summary of Units

Entity	Symbol for entity	Name of unit	Symbol for unit
magnetic flux	Φ	weber	Wb
magnetic flux density	B	tesla	T
Magnetomotive force	F	—	A
Permeability	μ	henry per metre	H/m
Reluctance	S	—	1/H

The long Solenoid

If a solenoid is long in comparison with the diameter of a turn it may be possible to assume that the whole of the m.m.f. is utilised in producing the flux inside the solenoid. If the length is l, and cross-section a, and if there is no magnetic core, then the reluctance is

$$l/a\mu_0$$

With a coil of N turns carrying current I, figure 10.11, the flux is therefore given by

$$\Phi = NI/(l/a\mu_0)$$
$$= NIa\mu_0/l \text{ Wb}$$
$$\therefore B = \Phi/a = NI\mu_0/l$$

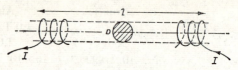

Figure 10.11   Long solenoid.

Example 10.5.   What current must flow in a narrow solenoid of length 1 m; number of turns 1000, in order that the magnetic field strength inside may be 1.8×10^{-5} T?

$$B = NI\mu_0/l$$
$$\therefore\ I = Bl/\mu_0\, N$$
$$= (1.8 \times 10^{-5} \times 1) \div (4\pi \times 10^{-7} \times 10^3)$$
$$= (1.8/4\pi) \times 10^{-1}\ \text{A} = 1.43 \times 10^{-2}\ \text{A}$$

If the solenoid contains a core of magnetic material of cross-sectional area a, and relative permeability μ_r, then the above expressions become

$$\Phi = NIa\mu_0\mu_r/l\ \text{Wb}$$
$$B = NI\mu_0\mu_r/l\ \text{T}$$

11 INDUCED E.M.F.

Dynamically induced e.m.f.

If a conductor is moved in a magnetic field in such a way as to cut across the lines of force of the field, an e.m.f. will be induced. If the conductor lies in the plane of the field or if it is moved along the lines of force, then no e.m.f. will be induced. Figure 11.1 shows two important cases. In figure 11.1a the conductor moves at right-angles

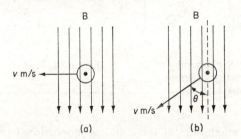

Figure 11.1 Dynamically induced e.m.f.

to the field. If the field strength is B, the conductor length l, and velocity v, all in SI units, then the induced e.m.f. is

$$E = Blv \text{ volts}$$

If the direction of motion is at an angle θ to the direction of the field, as in figure 11.1b, then

$$E = Blv \sin \theta \text{ volts}$$

The direction of the induced e.m.f. is given by a right-hand rule similar to the left-hand rule used before. It is illustrated in figure 11.2, the directions of pointing being

First finger   Field
Thumb   Motion
Second finger   Induced e.m.f.

Applying this rule to the case of figure 11.1 we see that the induced e.m.f. acts towards us.

The law governing the magnitude of the induced e.m.f. was discovered by Faraday; it is as follows: when an e.m.f. is induced in a circuit by a change in the number of lines of force through the

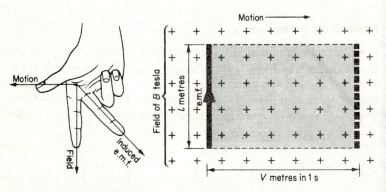

Figure 11.2   Right-hand rule. The directions correspond to Figure 11.1.

Figure 11. 3.

circuit, the magnitude of the e.m.f. is proportional to the rate of change of that number. In figure 11.3 the crosses indicate the lines of force of a field acting at right-angles to the paper. A conductor of length l moves with velocity v from left to right. In one second it sweeps out the shaded area, the number of lines cut therefore being Blv. But as these are cut in one second, Blv is also the rate of cutting. Also if the conductor is part of a closed circuit Blv is also the rate at which the lines of force thread or 'link' the circuit.

Example 11.1.   A small motor has two poles. The active length of each conductor on the armature is 8 cm and the diameter of the circle in which the conductors rotate 8.5 cm. The average field strength in the air-gap is 0.45 T, and the armature speed is 2000 rev/min.

What is the e.m.f. induced in each conductor?

Armature circumference $d = \pi \times 8.5 = 26.69$ cm
$$= 0.2669 \text{ m}$$
Rev/s $\quad = 2000/60 = 33.3$
∴ Peripheral speed $\quad = 0.2669 \times 33.3 = 8.88$ m/s
$B = 0.45$ T and $l = 8$ cm $= 0.08$ m
∴ $E = 0.45 \times 0.08 \times 8.88$
$$= 0.32 \text{ V}$$

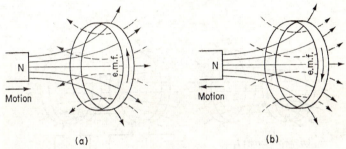

Figure 11.4 Induction in a circular coil. (a) Magnet approaching. (b) Magnet receding.

Statically induced e.m.f.

A statically induced e.m.f. is one which is produced when there is a change in the number of lines of force linking the circuit, when this change does not necessitate movement of the circuit relative to the field. It must be appreciated that there is no physical difference between dynamically and statically induced e.m.f.s since both obey Faraday's law. Figure 11.4a shows one way of generating a statically induced e.m.f., by moving a magnet whose lines of force link the coil. Figure 11.4a shows a N pole approaching the number of lines through the coil therefore increasing. Lenz's law states that if a current flows as the result of an induced e.m.f. this current will be in such a direction as to oppose the change producing it. We have seen that if a current flows round a flat coil one face of the coil will have N polarity and the other face S polarity. Hence it can oppose the change by producing its own magnetic flux in such a direction as to oppose the flux of the magnet when it is approaching, as in figure 11.4a, and to assist the flux of the magnet when it is receding, figure 11.4b. Applying the previous rules we see that the induced e.m.f. is counterclockwise when the coil is looked at from the left-hand side in case (a) and clockwise in case (b). As the coil figured is a closed circuit, induced currents in these directions will be set up in the two cases.

In figures 11.5a and b, the magnet is replaced by a coil which we will call the primary, P. The coil in which the e.m.f. is induced is the secondary, S. The current in P is such as to give a flux in the same direction as the magnet in figures 11.4a and b. In figure 11.5a the

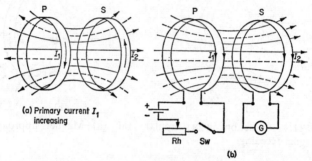

Figure 11.5   Production of a mutually induced e.m.f.

primary current is increasing, the flux, from left to right, through S is increasing and therefore the induced e.m.f. acts in a counter-clockwise direction when viewed from the left. As the circuit of S is closed a secondary current flows and, in accordance with Lenz's law it sets up a magnetic flux from right to left. Hence when the primary current is increasing, or is switched on, the induced current is in an opposite direction to the inducing current.

In figure 11.5b the primary current is decreasing and, the secondary coil circuit being closed, an e.m.f. and current are induced in such a direction that the secondary flux is now in the same direction as the primary flux. Hence the secondary induced current is now in the same direction as the inducing current.

In figure 11.5b suitable circuits for investigating these phenomena are shown. Coil S is connected to a galvanometer G which indicates the presence of a current and also its direction. Coil P is supplied from a cell (a Leclanché is suitable) and controlled by a switch Sw and variable rheostat Rh. By opening and closing Sw the effect of starting and stopping the primary current can be demonstrated,

Coefficient of Mutual Inductance, or Mutual Inductance

In figures 11.5a and b above the e.m.f. is induced in coil S by changes in the current in a neighbouring coil P. Such an e.m.f. is said to be mutually induced. In coil S we have

$$E_2 \text{ (per turn)} = \text{(Rate of change of flux threading the turn)}$$
$$\therefore E_2 \text{ for a coil of } N_2 \text{ turns} = N_2 \times \text{(Rate of change of flux)}$$

Suppose that coil S receives $1/m$ of the flux produced by coil P

$$\text{Then flux through } S = 1/m \times \text{(Flux due to P)}$$

If there is no iron in the magnetic circuit, or if we can assume that there is a straight-line law between Φ and I, then, flux due to P

$$\Phi = \binom{\text{Flux per}}{\text{amp in P}} \times I_1$$

$$\therefore \text{ Flux through } S = \binom{\text{Flux per}}{\text{amp in P}} \times I_1/m$$

$$\therefore \binom{\text{Rate of change of}}{\text{flux through S}} = \binom{\text{Flux per}}{\text{amp in P}} \times \binom{\text{Rate of}}{\text{change of } I_1}/m$$

$$\therefore E_2 = \left[(1/m) \times N_2 \times \binom{\text{Flux per}}{\text{amp in P}} \right] \times \binom{\text{Rate of}}{\text{change of } I_1}$$

Now the quantity in the square brackets is a constant for a given pair of coils. It is called the mutual inductance of coil S with respect to coil P, symbol $M_{2.1}$.

$$\therefore \binom{\text{Mutually induced}}{\text{e.m.f. in S}} = M_{2.1} \times \binom{\text{Rate of change of}}{\text{current in P}}$$

If the current in P changes at the rate of one A/s then the mutually induced e.m.f. in S is numerically equal to $M_{2.1}$. Hence the mutual inductance of a coil S with respect to a coil P is equal to the e.m.f. induced in S when the current in P changes at the rate of one ampere per second. The unit is the henry: symbol for the unit, H.

Self Induction

If the current in an isolated coil changes then the flux through it will change and Lenz's law will apply. If the current is increasing then the induced e.m.f. will act in opposition to the increase in flux and therefore to the increase in current. If the current is decreasing then it will be in such a direction as to augment the applied p.d. and thereby tend to keep the current at a steady value.

In this case it is easier to think of the two voltages, V the applied

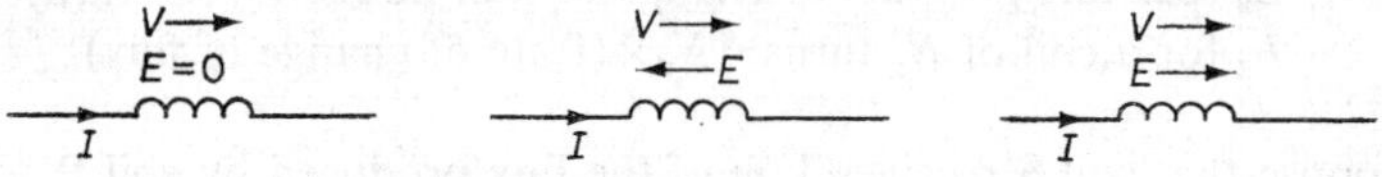

(a) Current steady   (b) Current increasing   (c) Current decreasing

Figure 11.6   Self-induced e.m.f.

p.d., and E the self-induced e.m.f. The three possible cases are illustrated in figure 11.6. The coefficient of self-induction or the *self inductance* is calculated as for the mutual inductance.

$$\text{Self-induced e.m.f.} = (\text{No. of turns}) \times \left(\begin{array}{c} \text{Rate of change} \\ \text{of flux} \end{array} \right)$$

Again, $\text{Flux} = (\text{Flux per amp}) \times \text{current}$

$$\therefore \left(\begin{array}{c} \text{Rate of change} \\ \text{of flux} \end{array} \right) = (\text{Flux per amp}) \times \left(\begin{array}{c} \text{Rate of change} \\ \text{of current} \end{array} \right)$$

$$\therefore \text{Self-induced e.m.f.} = \left[(\text{No of turns}) \times \left(\begin{array}{c} \text{Flux per} \\ \text{amp} \end{array} \right) \right] \times \left(\begin{array}{c} \text{Rate of change} \\ \text{of current} \end{array} \right)$$

The quantity in the square brackets is the self-inductance. It is denoted by the symbol L. The unit of self-inductance is the same as the unit of mutual inductance, namely the henry.

$$\therefore \text{ self-induced e.m.f.} = L \times \left(\begin{array}{c} \text{Rate of change} \\ \text{of current} \end{array} \right)$$

If the current changes at the rate of one ampere per second then the self-induced e.m.f. is numerically equal to L.

We can now understand the si unit of magnetic flux given in chapter 10. The si unit of electric inductance, whether self or mutual, is as follows:

The henry is the inductance of a closed circuit in which an electromotive force of one volt is produced when the current in the circuit varies uniformly at the rate of one ampere per second.

Example 11.2.   A lifting magnet produces a flux of 0.5 Wb. Its magnetising coil has 200 turns. If the circuit is suddenly opened when the full current is flowing, and the current takes 0.01 s to fall to zero, calculate the average value of the self-induced e.m.f. in the coil.

$$E_{\mathrm{av}} = \frac{\text{Change of flux}}{\text{Time of change}} \times \text{No. of turns}$$
$$= \frac{0.5}{0.01} \times 200 = 10\ 000 \text{ V}$$

This example shows that dangerously high voltages may be set up by the interruption of current in highly inductive circuits, and in practice, it is often necessary to take special precautions against them. The sparking seen on operating quite small power circuits is due to the same cause; the induced voltage being sufficient to break down the gap between the contact at the moment of opening or closing.

The RL series circuit

This is a circuit possessing both resistance and inductance in series. The behaviour of this circuit can be understood by reference to an easily understood mechanical analogy. The analogue of electrical resistance is frictional resistance since it limits velocity. The analogue of inductance is mass, or inertia, since it opposes change in velocity just as inductance opposes change in current.

Figure 11.7a shows a circuit having resistance only; the current rises to the Ohm's law value of V/R instantaneously.

Figure 11.7b shows a mechanical analogue, a truck without mass but with frictional resistance, μ. This can be defined as force divided by velocity when velocity is steady. Hence the truck reaches the velocity $v = F/\mu$ instantaneously.

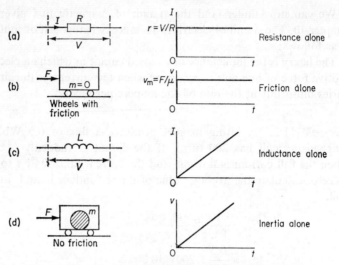

Figure 11.7   Mechanical analogues of electrical systems.

Figure 11.7c shows a circuit having inductance only. With $R = 0$ the applied p.d. has only to overcome the induced e.m.f.

$$\therefore \; V = L \times (\text{rate of change of current})$$

But V and L are constant, and therefore the rate of change of current is constant. This means that the current is proportional to time assuming the switch is closed at zero time so that the graph of current is a straight line through the origin. Again, the slope of this line is the rate of change of current and is therefore equal to V/L A/s. Similarly with the truck having mass m, but no resistance. The force F is used entirely in producing acceleration, a, and from Newton's second law.

$$F = ma$$
$$\therefore \; a = F/m, \text{ a constant}$$

Since acceleration is rate of change of velocity the velocity is proportional to time, giving the straight line through the origin as in figure 11.7d.

Now let the truck have both friction and inertia, figure 11.8. The force F now has two things to do: overcome the force of friction and produce the acceleration ma. At first v is zero and therefore $F = ma$, the acceleration being the same as in case 11.7d. The curve therefore starts off as the dashed straight line of figure 11.8. When v is finite and is gradually increasing, the component of F required for acceleration becomes progressively smaller, which means that the acceleration becomes progressively smaller. Eventually, the acceleration becomes negligibly small and the velocity therefore constant. This value v_m is that of figure 11.7b. Hence there is a gradual increase in velocity up to the value $v_m = F/\mu$, as shown by figure 11.8.

The case of the RL circuit is analogous to the above. At first with $i = 0$ (the symbol i being used for instantaneous values), the whole of the applied p.d. V is utilised in overcoming the self-induced e.m.f.

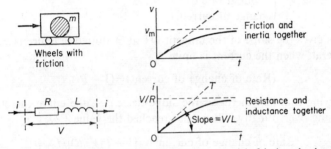

Figure 11.8 Mechanical and electrical systems with friction (resistance) and inertia (inductance).

so that, at first, the curve current rises from the origin at the same rate as in figure 11.7c. Its tangent at O is therefore the straight line of slope V/L. As the current increases there is less of the applied p.d. V available to overcome the induced e.m.f. which, in consequence, diminishes and therefore the rate of increase of current diminishes. The curve therefore turns over and eventually the current attains the value $I = V/R$.

Since the induced e.m.f. is always an opposing e.m.f., it is written with a negative sign, thus

$$\text{Self-induced e.m.f.} = -L \times (\text{rate of change of current})$$

$$\left.\begin{array}{c}\text{Mutually induced e.m.f.}\\ \text{in secondary}\end{array}\right\} = -M_{2.1} \times \left\{\begin{array}{c}\text{rate of change of}\\ \text{current in primary}\end{array}\right.$$

The voltage V applied to an RL circuit has to supply the ohmic drop Ri, where i is the current at any instant, and to overcome the induced e.m.f. of the inductor. Since this e.m.f. is considered to be negative because it opposes any change in the current, the component of V which has to overcome it must be both equal and opposite to it, and is therefore reckoned positive. The voltage equation thus becomes

Applied p.d. = ohmic volt-drop plus self-induced e.m.f.
$$V = Ri + L \times (\text{rate of change of current})$$

At the moment of application of V, that is, at the instant $t=0$, $i=0$ and therefore, at this instant

$$V = L \times \text{initial rate of change of current}$$

$$\therefore \left(\begin{array}{c} \text{Initial rate of} \\ \text{change of current} \end{array} \right) = V/L \text{ A/s}$$

This gives the slope of the tangent OT at O, dotted in figure 11.8. In general, when the current is finite

$$(\text{Rate of change of current}) = (V - Ri)/L$$

showing that as i increases the slope decreases, the curve therefore turning over. Finally, when i has reached the value V/R

$$(\text{Rate of change of current}) = [V - R(V/R)]/L = 0$$

the ultimate graph of current therefore bring the horizontal straight line of ordinate V/R.

It can be shown that the current rises to one half of the maximum value of V/R in $0.693\ \tau$ s, where $\tau = L/R$. This is called the time constant.

Example 11.3. A p.d. of 100 V is applied to a RL series circuit having $R = 5\ \Omega$ and $L = 0.2$ H. Plot the curve of current against time.

$$V/L = 100/0.2 = 500 \text{ A/s}$$

The tangent OT therefore corresponds to a rate of rise of 500 A/s. Thus for $I = 20$ A, $t = 20/500 = 0.04$ s. This locates the point T, figure 11.9, so that the tangent OT can be drawn.

The final value of the current $I = 100/5 = 20$ A.

The current rises to one half of the maximum value in $0.693\ L/R$ s. Hence the current will reach 10 A in

$$0.693 \times (0.2/5) = 0.693 \times 0.04 = 0.0267 \text{ s}$$

This gives point P at (0.0267 s, 10 A).

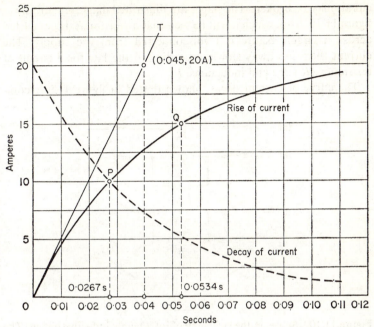

Figure 11.9 Rise and decay of current in a *RL* circuit for which $V=$ 100 V, $R=5\,\Omega$, $L=0.2$ H.

It can also be shown that the current rises to one-half of the remaining possible rise in a further time of 0.693 L/R, namely 0.0267 s in this case. One half of the remaining possible rise is 5 A, bringing the current to 15 A when $t=0.0534$ s. This gives point Q at (0.0534 s, 15 A). Further points can be obtained in the same way if desired.

We see that when a *RL* current is switched on to a steady d.c. supply, the circuit behaves at the moment of switching on as though it possesses inductance only. After sufficient time has elapsed for the current to become steady it behaves as though it possesses resistance only. If, while carrying full current the circuit is closed on itself and the p.d. removed the current dies away gradually to zero. This is analogous to removing the impressed force from the truck with both friction and inertia after it has attained full velocity: it gradually comes to rest. In the closed circuit there is no dangerous rise in voltage, and this indicates how the circuit of a highly inductive circuit may be broken. The stages are illustrated diagrammatically in figure 11.10, from which it will be seen that a resistor R' is connected across a circuit before it is disconnected from the supply. The current then dies away slowly according to the law for a circuit of resistance $(R+R')$ and inductance L.

The curve of current decay, shown dotted in figure 11.9 is com-

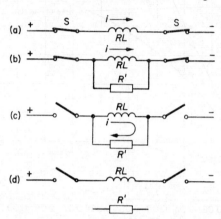

Figure 11.10   Stages in the switching off of a highly inductive circuit. The two poles, S, of the main switch are shown separated, to simplify the diagram.

- (a) Normal operation.
- (b) Inductor shunted by resistor R'.
- (c) Current decays according to the law of a $(R+R')L$ circuit.
- (d) Isolation.

plimentary to the curve of current rise. The sum of the ordinates is a constant and equal to V/R.

Energy of the Magnetic Field

If N and S poles of two separate magnets are in contact force will be required to separate them and consequently work will be done. Since energy cannot be destroyed, the energy expended in separating the poles will be converted into some other form. In this case it will be in potential energy stored in the magnetic field created by the separation of the poles.

Example 11.4.   A 300 kW, 500 V, shunt generator has 8 poles, the flux per pole being 7.3×10^{-2} Wb. The self induction of each magnetising pole is 13.65 H and the field current is 6.95 A. Calculate (a) the total stored energy in the magnetic circuit of the machine and (b) the e.m.f. which would be induced if the field circuit was suddenly opened while carrying this current.

$$L = 8 \times 13.65 = 109.2 \text{ H}$$
$$\therefore \text{ stored energy } W = \tfrac{1}{2} LI^2 \dagger$$
$$= \tfrac{1}{2} \times 109.2 \times (6.95)^2$$
$$= 2630 \text{ J}$$

If the current falls to zero in 0.01 s, then rate of charge of current

$$6.95/0.01 = 695 \text{ A/s}$$
$$\therefore \text{ self induced e.m.f.} = 109.2 \times 695 \simeq 76\,000 \text{ V}$$

Such a voltage would wreck the insulation of the machine and it explains why in practice the field circuits are never opened when carrying current.

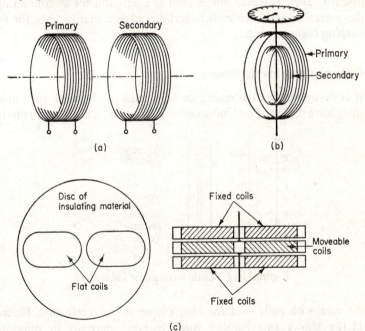

Figure 11.11 Three types of variable inductor.

† This is analogous to the expression $\tfrac{1}{2}CV^2$ for the stored energy in an electric field. It is derived on pages 124–6.

Variable Inductors

Variable inductors, both self and mutual, are used for laboratory purposes. Figure 11.11 shows three forms, figure 11.11a is a very simple form consisting of two coaxial coils whose distance apart can be varied and measured on a scale. The coils can also be connected so that the currents are in the same or opposite directions. If the two coils are connected in separate circuits the device becomes a variable mutual inductor. In figure 11.11b one coil can rotate inside the other, the inductance depending on the angle between the planes of the coils. Since a rotation of 180° gives a reversal of the field of the inner coil, it is not necessary to arrange for a reversal of the current in one of the coils. Types (a) and (b) can be very easily made. Figure 11.11c shows a more sophisticated instrument. It consists of three flat discs of insulating material each having two flat coils. They are arranged in sandwich form, the outer discs being fixed and the inner disc rotatable. This also can be used as a self inductor by connecting the moveable and fixed coils in series, and as a mutual inductor by keeping them separate.

Coils with very low inductance

It is desirable that the resistance coils used in electrical apparatus shall have as low a self-inductance as possible. This is particularly

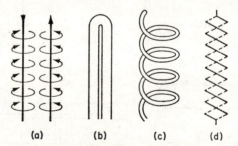

(a)   (b)   (c)   (d)

Figure 11.12   Coils having low inductance.

the case with coils used for high frequency measurements. Figure 11.12a shows two parallel wires carrying currents in opposite directions, their fields therefore being in opposite directions. If they could be brought so intimately together that both occupy the same space the total field would be zero. Actually the best that can be done is to make the wires touch one another as in Figure 11.12b. This practically eliminates the external field except very close to the wires, but the fields inside the wires remain, and therefore zero inductance is an impossibility.

Thus, by bending back a length of wire on itself the inductance can be made very low, but not zero. If this doubled wire is wound in a coil as in figure 11.12c the inductance will be considerably less than that of a simple wire coil of the same dimensions and turns. This type of coil is generally used in resistance boxes. Another method is to wind two coils on the same former, one right-handed, the other left-handed, and to connect them in parallel as in figure 11.12d.

Summary of Units

Name of entity	Symbol for entity	Name of unit	Symbol for unit
Self inductance	L	henry	H
Mutual inductance	M	henry	H

12 CAPACITORS INSULATING MATERIALS

Figure 12.1 shows a reservoir of very large capacity, supplying a small vessel via a pipe. If the head of water in the vessel is less than that in the reservoir, a current of water will flow along the pipe to the vessel, but as soon as the head of water is the same as that in the reservoir, this flow of water ceases. The filling of the vessel is thus accomplished by a current of temporary nature. Let E be the constant head of water in the reservoir and V the head in the vessel

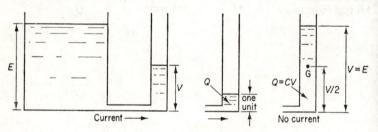

Figure 12.1   Hydraulic analogy.

at any instant, then current in the pipe will flow as long as $E > V$ and will cease when $E = V$.

Let the capacity of the vessel be reckoned, not by the amount of water required to fill it when poured in from the top, but by the amount from the reservoir required to make V equal to unity. Represent this by C. Then when the vessel is filled and V has its maximum value $V = E$, the quantity of water in it is CV. Suppose the system is such that CV is the weight of water, then since the centre of gravity G is at a height of $V/2$

$$\text{Potential energy} = (V/2) \times VC$$
$$= \tfrac{1}{2} CV^2$$

124

The above simple hydraulic system is the analogue of the electric capacitor which is simply two conductors insulated from one another, figure 12.2, which shows a capacitor consisting of two parallel metal plates joined to the terminals of a battery of e.m.f. E. The plates will acquire the potentials of the terminals to which they are connected and to do this electrons will flow to the negative plate B, thereby creating a p.d. which, at any instant, we represent by V. As long as $V > E$ this electron flow will take place, this being equivalent to a conventional current i out at the $+$ terminal of the battery and in at the $-$ terminal. This current does not flow between the capacitor

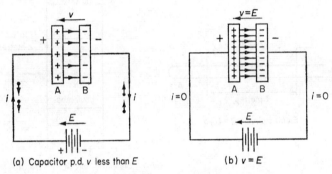

(a) Capacitor p.d. v less than E (b) $v = E$

Figure 12.2   Charging of a capacitor.

plates, its function being to supply electrons to plate B. It is therefore called a displacement current. When sufficient electrons have been displaced to make the p.d. equal to E the displacement current ceases.

It will be seen that the capacitor p.d. is in opposition to E as far as the battery circuit is concerned and therefore energy must be expended by the battery in order that the displacement current may flow. This energy is converted to the potential energy of the electric field which is created between the plates.

Let C equal the quantity of electricity in coulombs required to raise the p.d. between the plates by 1 V. Then C is the capacity of the capacitor. The unit is clearly the coulomb per volt. The SI unit of capacitance is called the Farad, symbol F, and is defined as follows.

The farad is the capacitance of a capacitor between the plates of which there appears a difference of potential of one volt when it is charged by a quantity of electricity equal to one coulomb.

By analogy with the hydraulic system of figure 12.1 for which we used corresponding symbols we have for the energy stored in the electric field between the plates

$$W = \tfrac{1}{2}CV^2$$
$$Q = CV$$
$$W = \tfrac{1}{2}QV = \tfrac{1}{2}Q^2/C$$

The unit of energy is the joule.

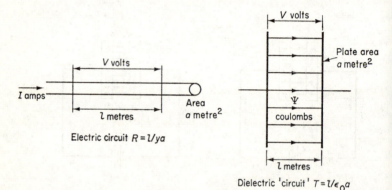

Figure 12.3   Analogy between electric and dielectric circuits.

Capacitance of a parallel plate capacitor

We will assume, first of all, that the medium between the plates is air. In the SI system the total electric flux Φ is equal to the quantity Q at the terminations of the lines of force of the electric field.

$$\therefore \quad C = \Phi/V \ \mathrm{AV^{-1}}$$
$$\text{or} \quad \Phi = CV \text{ coulombs}$$

We will now introduce a new concept T, the elastance, where $T = 1/C$ the reciprocal of the capacitance. We then have

$$\Phi = V/T \text{ coulombs}$$

In this form there is a similarity to the fundamental equations for the electric current circuit and the electric 'field' circuit. Thus from figure 12.3

$$R = l/\sigma a \text{ and } T = l/\epsilon_0 a$$

Where σ is the conductivity in the current case and ϵ_0 is the electric space constant in the dielectric case.

Example 12.1.   A capacitor consists of two parallel plates in air, each of area 2 m², and their distance apart 0.5 cm. If the capacitance is 3.54×10^{-9} F, what quanitity of electricity is displaced when the p.d. between the plates is made equal to 10 000 V?

$$C = 3.54 \times 10^{-9} \text{ F}, \ V = 10\ 000$$
$$\therefore \ Q = 3.54 \times 10^{-9} \times 10^4$$
$$= 3.54 \times 10^{-5} \text{ C}$$

Example 12.2.   Verify the stated value of the capacitance in example 12.1 given that $\epsilon_0 = 8.85 \times 10^{-12}$.

For the distance between the plates we will use t instead of l, as this is more general. Then

$$a = 2 \text{ m}^2, \ t = 0.5 \text{ cm} = 5 \times 10^{-3} \text{ m}$$
$$C = 1/T = \epsilon_0 a/t$$
$$= 8.85 \times 10^{-12} \times 2/(5 \times 10^{-3})$$
$$= 3.54 \times 10^{-9} \text{ F}$$

This example shows that the farad is a very large unit for which reason capacitances are often expressed in microfarads, μF, picofarads, pF. Thus in the above example

$$C = 3.54 \times 10^{-9} \text{ F} = 3540 \times 10^{-12} \text{ F} = 3540 \text{ pF}$$

Relative Permittivity

If a dielectric such as glass, paraffin wax, mica, etc., is placed between the plates of a capacitor the capacity is increased. If the dielectric completely fills the space between the plates the ratio of the new capacitance to the old with air (or more strictly speaking, a vacuum) as the medium is called the relative permittivity ϵ_r of the medium. Thus the general expression for the capacitance of a parallel plate capacitor is

$$C = \epsilon_0 \epsilon_r a/l$$

A few values are given below

Medium	Relative Permittivity
Air	1.00059
Glass	4 to 10
Ebonite	2.8
Paraffin wax	2.2
Mica	4 to 8

Example 12.3. A parallel plate capacitor has plates of 3m² area spaced 2.5 mm apart. The space between the plates is filled by a dielectric for which $\epsilon_r = 3$. Calculate the capacitance, the charge and the stored energy when the p.d. between the plates is 10 000 V.

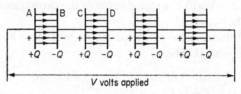

Figure 12.4   Capacitors in series.

$$C = \epsilon_0 \epsilon_r a / l = 8.85 \times 10^{-12} \times 3/(2.5 \times 10^{-3})$$
$$= 10.62 \times 10^{-9} \text{ F}$$
$$= 10.62 \times 10^{-3} \mu\text{F or } 10\,620 \text{ pF}$$
$$Q = CV$$
$$= 10.62 \times 10^{-9} \times 10^4$$
$$= 10.62 \times 10^{-5} \text{ C}$$
$$W = \tfrac{1}{2}CV^2$$
$$= \tfrac{1}{2} \times 10.62 \times 10^{-9} \times 10^8$$
$$= 5.31 \times 10^{-1} \text{ J}$$

Capacitors in Series

Figure 12.4 shows a number of capacitors in series. When the displacement current has died down to zero the p.d. between the first positive plate and the last negative will be equal to V, the supply p.d. Tubes of force starting on plate A will terminate on plate B so that, if $+Q$ is the charge of plate A, $-Q$ will be the charge on B. Before charging, plates B and C formed part of a single unelectrified conductor with no surplus of either positive or negative electricity. Hence when B acquires a charge $-Q$, C must acquire a charge $+Q$; and so on for the complete series. Hence all the capacitors, whatever their individual capacitances, have equal charges. Calling the total capacitance C we have

$$V = V_1 + V_2 + V_2 + ...$$
$$\therefore \quad \frac{Q}{C} = \frac{Q}{C_1} + \frac{Q}{C_2} + \frac{Q}{C_3} + ...$$
$$1/C = 1/C_1 + 1/C_2 + 1/C_3 + ...$$
$$\text{or} \quad T = T_1 + T_2 + T_3 + ...$$

The way the applied p.d. V, divides between the various capacitors is conveniently determined by comparison with an electric circuit,

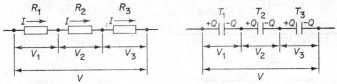

Figure 12.5   Capacitors in series compared with resistors in series.

having a number of resistors in series (figure 12.5). The comparison is as follows

$$R = R_1 + R_2 + R_3 + ... \qquad T = T_1 + T_2 + T_3 + ...$$
$$I = V/R \qquad\qquad\quad Q = V/T$$
$$V_1 = IR_1 \qquad\qquad\quad V_1 = QT_1$$
$$V_2 = IR_2 \qquad\qquad\quad V_2 = QT_2$$
$$V_2 = IR_3 \qquad\qquad\quad V_2 = QT_3$$
$$\text{etc.} \qquad\qquad\qquad\quad \text{etc.}$$

Example 12.4 (a) A p.d. of 100 V is applied to three resistors of 1, 2 and 3 Ω in series. Calculate the current and the p.d. across each resistor.

$$R = 1 + 2 + 3 = 6\ \Omega$$
$$I = V/R = 100/6 = 16.67\ \text{A}$$
$$V_1 = IR_1 = 16.67 \times 1 = 16.67\ \text{V}$$
$$V_2 = IR_2 = 16.67 \times 2 = 33.34\ \text{V}$$
$$V_3 = IR_3 = 16.67 \times 3 = 50.01\ \text{V}$$

(b) A p.d. of 100 V is applied to three capacitors of 1, 2 and 3 μF in series. Calculate the quantity in each and the p.d. across each.

$$T_1 = 1/C_1 = 1/10^{-6} = 10^6$$
$$T_2 = 1/C_2 = 1/2 \times 10^{-6} = 0.5 \times 10^6$$
$$T_3 = 1/C_3 = 1/3 \times 10^{-6} = 0.333 \times 10^6$$
$$T = (1 + 0.5 + 0.333) \times 10^6 = 1.833 \times 10^6$$
$$Q = V/T = 100/(1.833 \times 10^6) = 5.46 \times 10^{-5} \text{ C}$$
$$V_1 = QT_1 = 5.46 \times 10^{-5} \times 10^6 = 54.6 \text{ V}$$
$$V_2 = QT_2 = 5.46 \times 10^{-5} \times 0.5 \times 10^6 = 27.3 \text{ V}$$
$$V_3 = QT_3 = 5.46 \times 10^{-5} \times 0.333 \times 10^6 = 18.2 \text{ V}$$

Capacitors in Parallel

The corresponding electric circuit arrangement to resistors in

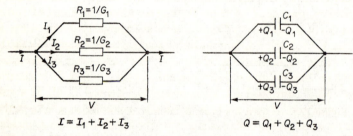

$$I = I_1 + I_2 + I_3 \qquad\qquad Q = Q_1 + Q_2 + Q_3$$

Figure 12.6   Capacitors in parallel compared with resistors in parallel.

parallel is given in figure 12.6. G is the conductance, the reciprocal of R. The comparison is as follows

$$I = I_1 + I_2 + I_3 + \ldots \qquad\qquad Q = Q_1 + Q_2 + Q_3 + \ldots$$
$$I_1 = V/R_1 = VG_1 \qquad\qquad Q_1 = VC_1$$
$$I_2 = V/R_2 = VG_2 \text{ etc.} \qquad\qquad Q_2 = VC_2 \text{ etc.}$$
$$\therefore \quad I = VG_1 + VG_2 + VG_3 + \ldots \quad \therefore \quad Q = VC_1 + VC_2 + VC_3 + \ldots$$
$$\text{But } I = VG \qquad\qquad\qquad \text{But } Q = VC$$
$$\therefore \quad G = G_1 + G_2 + G_3 + \ldots \qquad \therefore \quad C = C_1 + C_2 + C_3 + \ldots$$

We thus have the following rules: with capacitors in series, the total elastance is the sum of the separate elastances. With capacitors in parallel the total capacitance is the sum of the separate capacitances.

Example 12.5. Three capacitors of 1, 2 and 3 μF are in parallel. If they are connected to a 100 V battery what will be the total quantity of electricity stored?

$$C = C_1 + C_2 + C_3 = 1 + 2 + 3 = 6\mu F = 6 \times 10^{-6} \text{ F}$$
$$Q = CV$$
$$= 6 \times 10^{-6} \times 100 = 6 \times 10^{-4} \text{ C}$$

Dielectric Hysteresis

This is the loss of energy experienced by dielectrics when they are subjected to alternating electric stress. Many dielectric molecules are very complex in the arrangement of the electrons and protons, so that the 'centre of gravity' of the negative charges is not in the same place as that of the positive charges. Such molecules are therefore electric dipoles, analogous to magnetic dipoles. In an unstressed dielectric the orientations of these dipoles are random, but when under electric stress, as for example between the plates of a charged capacitor, they tend to align themselves along the lines of force. This involves the expenditure of energy. Similarly if the stress is reduced the dipoles do not make the previous movements in exactly the reverse sense with the result that, after one cycle of electrical stress the expenditure of energy is finite. This is called the dielectric loss and if continued for an appreciable time will result in a rise in temperature. There are some dielectric molecules in which, in the unstressed state, the centre of gravity of the negative charges coincides with that of the positive charges. Under electric stress these are pulled apart, the molecules thereby becoming induced dipoles.

There is no such energy expenditure when the dielectric is air.

Dielectric Materials

The following are the important electrical properties of dielectrics:
1. High resistivity.
2. High breakdown strength.
3. Relative permittivity greater than unity.
4. Dielectric loss when subjected to alternating electric stress.

No. 4 is of little importance with low voltage appliances; it becomes important at voltages of 66 000 or more. It is most important in extra-high voltage cables because, with these, there are, as a rule, no paths of good heat conductivity to enable the heat generated by the cable losses, of which the dielectric loss is one, to be dissipated.

Other properties are also essential, depending on the use to which the material is put. For materials used in the manufacture of electrical machines the following are essential:

1. Sufficient mechanical strength to withstand vibration, and the bending and abrasion experienced during manufacture.
2. Good heat conductivity.
3. Ability to withstand the maximum working temperature without change in physical properties or chemical composition.
4. Non hygroscopic, unless the appliance can be sheathed to prevent the absorption of moisture.

Properties of Dielectric Materials

Material	Breakdown strength at 50 Hz kV mm⁻¹	Resistivity M Ω–cm	Relative permittivity	Affected by moisture	Safe temperature	Uses
Asbestos	3–4.5	1.6×10^5	—	absorbent	500° C	covering of wires in highly rated machines
Bakelite	20–25	—	5–6	no	200	bakelised paper made up in the form of boards. Many uses
Bitumen (vulcanised)	14	—	4.5	no	about 60	low-voltage mining cables; cable-box filling compound
Cotton	3–4	1000 upwards acc. to dryness	—	absorbent	90	covering for wires
Empire cloth	10–20	as cotton	2	absorbent if varnish layer is cracked	90	wrappings for wires arranged in groups; for example armature coils
Fibre	5	as cotton	4–6	ditto	90	in sheet form, slot linings
Mica	40–150	$5–100 \times 10^6$	3–8	no	500 or more	not generally used in pure form
Micanite	30	$10 \times 600 \times 10^6$	6–8	no	130 when under pressure	commutator segments. Slot linings for high voltage machines; bushes
Paper	4–10	as cotton	2	absorbent	90	cable insulation when oil impregnated; covering for transformer conductors
Paraffin wax	8	3×10^{10}	2	no	under 50	—
Porcelain	9.20	$1–1000 \times 10^6$	4–7	not when vitreous and glazed		insulators for overhead conductors
Rubber	10–25	$2–10 \times 18^8$	2–3	slightly	40	cable insulation

Practical Capacitors

The farad is such a vast unit that a capacitor with only two plates would have to be of impossible size to have a capacitance of one farad.

Example 12.6.   What would be the necessary plate area of a parallel

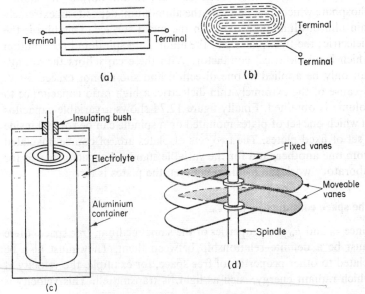

(a)

(b)

Terminal

Terminal

Terminal

Terminal

Insulating bush

Electrolyte

Aluminium
container

Fixed vanes

Moveable
vanes

Spindle

(c)

(d)

Figure 12.7   Various types of capacitor.

plate capacitor with two plates, separation 1 mm, to give a capacitance of one farad. Air dielectric.

$$\text{From } C = \epsilon_0 a/t$$
$$a = Ct/\epsilon_0 \quad (t = 10^{-3} \text{ m})$$
$$= 1 \times 10^{-3}/8.85 \times 10^{-12} = 1.13 \times 10^8 \text{ m}^2$$

Square plates would therefore have to be of 1.064×10^4 m or 6.6 miles square.

For this reason practical capacitors, apart from those whose capacitances are best expressed in pF, are of multi-plate construction. Figure 12.7 shows several types. In figure 12.7a metal sheets, often tinfoil, are separated from one another by paraffined paper in cheap construction, or mica for more expensive construction. Even-numbered plates are connected to one terminal and odd-numbered plates to the other. In another form, figure 12.7b, the plates consist of long strips of tinfoil separated by a thin flexible dielectric such as paraffined paper, rolled up spirally as shown, and the whole soaked in melted paraffin wax, which sets solid when cold. The electrolytic capacitor (figure 12.7c) consists of a spiral of aluminium sheet immersed in a solution such as ammonium borate or sodium phosphate which, reacting with the aluminium, deposits an extremely thin film of aluminium oxide on the metal surface. This film is the dielectric; the two 'plates' are the aluminium spiral and the solution which is an electrical conductor. With these capacitors the voltage can only be applied in one direction and should not exceed 50 V. Because of the extremely thin dielectric, a high ratio capacitance to volume is obtained. Finally figure 12.7d shows a variable capacitor in which one set of plates mounted on a spindle can be rotated inside a set of fixed plates. The two sets of plates are, of course, insulated from one another, and for the variable standard capacitors used for laboratory work the medium between the plates is air.

The space constants μ_0 and ϵ_0

Since ϵ_0 and μ_0 are properties of the same medium, free space, there must be a definite relationship between them. They must also be related to other properties of free space, for example, the velocity at which radiant energy, such as light, is transmitted. This velocity is

$$c = 3 \times 10^8 \text{ m/s}$$

The researches of Clark Maxwell led to the relationship

$$\epsilon_0 \mu_0 = 1/c^2 = 1/(9 \times 10^{16})$$
$$\therefore \quad \epsilon_0 = 1/(9 \times 10^{16} \times \mu_0)$$

The value of μ_0 depends on the system of units adopted as it is only the product $(\epsilon_0 \mu_0)$ which is a constant. In the SI system μ_0 has to be

$$\mu_0 = 4\pi \times 10^{-7}$$
$$\therefore \quad \epsilon_0 = 1/(9 \times 10^{16} \times 4\pi \times 10^{-7})$$
$$= 8.85 \times 10^{-12}$$

The reason for the presence of a velocity squared in the relationship between the electrostatic and magnetic systems of units can also be explained as follows: The force between two equal charges q, separated by a distance d is proportional to q^2/d^2. The force between two long parallel conductors each carrying I amperes, and distance d apart, is proportional to I^2/d. But a current is produced by the movement of electric charges and is given by the product of the charge q and its velocity v. The force between the two long parallel conductors thus becomes proportional to q^2v^2/d, showing that the

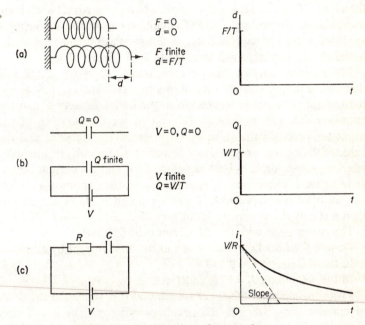

Figure 12.8   Charging of a capacitor.

coefficients of many of the formulae are related to velocity. Maxwell showed that this velocity is c, the velocity of light. Thus $\epsilon_0\mu_0 = 1/c^2$.

Charge and discharge of a capacitor

The mechanical analogue of a capacitor is a spring. If a force F is applied at the end of a spring, as in figure 12.8a, there is an extension d proportional to F. We can therefore write

$$F = Td \quad d = F/T$$

where T is the elasticity of the spring expressed as the extension due to unit force. This is analogous to the elastance of a capacitor, which is the quantity of electricity stored with unit applied p.d. Assuming that the spring has no mass and that there is no resistance to deflection, the displacement d is attained instantaneously. Similarly with a capacitor, if there is no resistance in the circuit and no loss of energy of any kind, the application of a p.d. V results in a charge of $Q = VC$ or V/T being acquired instantaneously (figure 12.8b). It therefore follows that if a p.d. is applied to a series circuit containing a capacitor, for example a RC circuit, then, at the moment of application the capacitor will act as though short-circuited, and the initial charging current will be equal to V/R (figure 12.8c).

The moment current flows the capacitor begins to charge and therefore a p.d. v_c is established across its terminals. The p.d. across the resistor is therefore reduced from the initial value of V and the current reduced from V/R to $(V - v_c)/R$. As the charging of the capacitor proceeds the value of v_c progressively increases and the value of the current progressively decreases. Eventually, theoretically after an infinite time, v_c becomes equal to V and the current ceases, as in figure 12.8c. We can say, at this stage, the capacitor now acts like an open circuit; in fact, if a series circuit contains a capacitor then a steady direct current is not possible.

The comparison with the RL circuit is as follows:

Denote the initial current by I_0 and the final current by I

In the RL circuit $I_0 = 0$ and $I = V/R$.

In the RC circuit $I_0 = V/R$ and $I = 0$.

From the shape of the curve we see that it is similar to the delay of current curve when a RL circuit carrying current is suddenly short-circuited on itself. The initial current is V/R; this falls to one-half in 0.693 τ seconds, where τ is the time constant; to one-quarter after another 0.693 τ seconds, and so on. We therefore need to know the time constant. It is CR s.

The proof is as follows:

When a capacitor is being charged the quantity of electricity q at any instant is $q = Cv_c$. Now C is a constant

$$\therefore \quad \text{Rate of charge of } q = C \times \text{rate of change of } v_c$$

Again, quantity is current multiplied by time and therefore the rate of change of q at any instant is equal to the current i at that instant

$$\therefore \quad i = C \times \text{rate of change of } v_c$$

The p.d. across the resistor when the capacitor p.d. is v_c is $(V - v_c)$

$$\therefore \; i = (V - v_c)/R$$
$$\therefore \; (V - v_c)/R = C \times \text{rate of change of } v_c$$
$$v_c = V - CR \times \text{rate of change of } v_c$$
$$V = v_c + CR \times \text{rate of change of } v_c$$

when $t = 0$, $v_c = 0$ and therefore at that instant

$$\text{Rate of change of } v_c = V/CR$$

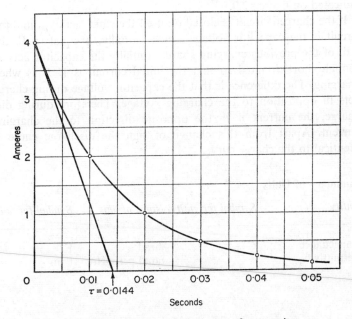

Figure 12.9   Charge and discharge of a capacitor.

Again, since $i = (V - v_c)/R$
and V and R are constant, the rate of change of $i = V/(R \times CR)$
$$= I_0/CR$$
By comparison with the equation for the RL circuit we see that the time constant is CR

Example 12.6.   An RC circuit for which $R = 250 \, \Omega$ and $\tau = 0.01443$ s has a p.d. of 1000 V applied. Plot the curves of charging and discharging current.

The initial current is $V/R = 1000/250 = 4$ A

The initial rate of change is obtained by drawing the tangent line through the points ($i=4$, $t=0$) and ($i=0$, $t=\tau=0.0144$), as shown in figure 12.9.

The current falls to one-half of 4, namely 2 A in 0.693 $\tau = 0.693 \times 0.01443 = 0.01$ s.

It falls to one-half of 2, namely 1 A in another 0.01 s. It falls to 0.5 A in another 0.01 s, and so on. These and subsequent points are indicated on figure 12.9.

If the charged circuit is closed on itself the e.m.f. acting round the circuit is the capacitor voltage which, initially, is equal to V, the p.d. of the previous charging source. Initially the capacitor acts as though short-circuited so that the initial current is V/R as when charging. The difference is that the capacitor voltage during charge acts in opposition to the charging voltage. Therefore during discharge, the current is in the opposite direction to the charging current. Apart from this change of sign the discharge curve is identical to the charge curve.

Summary of Units

Entity	Symbol for entity	Name of unit	Symbol for unit
Capacitance	C	farad	F
Permittivity	ϵ	farad per metre	F/m

13 CHEMICAL EFFECTS

Electrolysis

When current flows in a metallic conductor, such as copper, there is no change in the apparent state of the conductor, apart, perhaps, from a rise in temperature. This is because the current is the result of an axial drift of electrons and as many electrons enter at one end as leave at the other. There are many liquids, mainly solutions of ionic compounds in which the passage of current is accompanied by a chemical change. Such liquids are called electrolytes.

Consider a compound such as sodium chloride, the molecule of which is formed by the association of one sodium and one chlorine atom. Sodium has a single outermost electron while chlorine has a gap of one electron in its outermost shell so that, in a sense we can regard the sodium electron as looking for a gap to fill, and this gap is supplied by the chlorine atom. Thus the two come together to form the molecule NaCl, as shown in a somewhat fanciful manner in figure 13.1a. The total numbers of electrons and protons are equal so that the molecule is electrically neutral.

The attractive power in terms of the number of electronic charges is called the valency. Thus hydrogen and chlorine are monovalent, but oxygen is divalent. One atom of oxygen can therefore combine with two of hydrogen to form a molecule of water, H_2O (figure 13.1b).

Under certain circumstances the constituent atoms can dissociate. This takes place to a great extent in aqueous solution. We have seen that the attraction between two unlike charges is inversely proportional to the relative permittivity of the medium. For water this has the high value of 81, resulting in a great deal of dissociation in water solutions. In the case of sodium chloride there is some dissociation without the presence of water, and fused sodium chloride is an electrolyte.

139

Now consider the important sulphuric acid molecule, H_2SO_4. In the absence of water the structure is that of figure 13.1c. In a water solution each acid molecule reacts with two of water thus:

$$H_2SO_4 + 2H_2O = SO_4{}^{2-} + H_3O^+ + H_3O^+$$

Thus the SO_4 group is negatively charged with a charge of $2e$ while each H_3O group has a charge of $+e$. The $SO_4{}^{2-}$ group can exist in this state so long as it retains its negative charge; if it loses its charge then it must take place in some chemical action, to be discussed

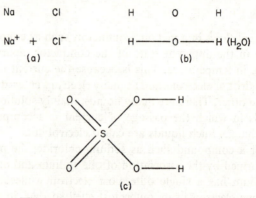

Figure 13.1   Structure of some molecules.

below. When copper sulphate, $CuSO_4$, is dissolved in water dissociation leads to the change

$$CuSO_4 = Cu^{2+} + SO_4{}^{2-}$$

Now let two plates, electrodes, be immersed in the solution and let these be connected to a source such as a battery. The positive electrode is called the anode and the negative the cathode. With dilute sulphuric acid the phenomena are represented in figure 13.2. The charged systems SO_4^{2-} and H_3O^+ are mobile, for which reason they are called ions. The SO_4^{2-} ions move to the anode and the H_3O^+ ions to the cathode. Since movements of electric charges constitute an electric current we see that in an electrolyte (a) the current is due, not to electrons but to charged assemblages of atoms such as SO_4^{2-}; (b). The current consists of two components; positive charges moving to the cathode and negative charges to the anode.

Action at the Electrodes

We will consider the end result only. When the SO_4^{2-} reaches the anode the two additional electrons are attracted to the positive anode

$$SO_4^{2-} = SO_4 + 2e^-$$

The electrons enter the anode and become part of the external current. The SO_4 radical without its charge can not exist in that form and must therefore enter into some kind of chemical combination.

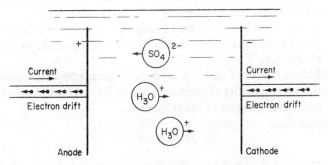

Figure 13.2   Electrolysis of dilute sulphuric acid.

If the anode is inert, such as platinum or carbon, then it reacts with the water

$$2SO_4 + 2H_2O = 2H_2SO_4 + O_2$$

Thus sulphuric acid is formed and oxygen gas given off. If the anode consists of a metal such as copper, then

$$SO_4 + Cu = CuSO_4$$

Thus copper sulphate is formed and enters into the solution. No gas is evolved.

The H_3O^+ ions move to the cathode. Now the positive charge is due, not to the presence of an extra positive charge but to the absence of a negative charge so that there is no question of the 'giving up' of the positive charge. The end effect, which is all we are concerned with, is that two electrons enter from the cathode

$$2H_3O^+ = 2H_3O + 2e^- = 2H_2O + H_2 + 2e^-$$

showing (a) that hydrogen gas is liberated at the cathode and (b) the number of electrons entering the external circuit at the anode is equal to the number leaving it at the cathode. Thus a steady state is maintained. We see from the above that the products of electrolysis are liberated only at the electrodes and that these products are formed only when a p.d. of sufficient magnitude is applied to the electrodes.

Faraday's Laws of Electrolysis

By his researches Faraday discovered that the phenomenon of electrolysis is governed by two laws
    1. The mass of a substance liberated at an electrode is proportional to the quantity of electricity which has passed through the electrolyte.
    2. The masses of different substances liberated by the passage of the same quantity of electricity are in the ratio of the chemical equivalent weights of the substances.
    Hence mass

$$M = kIt = kQ$$

The constant k, for any given atom, or radical, is a composite quantity depending on the atomic, or group, weight, and on the valency. These factors are taken into account in a constant called the electro-chemical equivalent, e.c.e. Denoting it by z we have

$$M = zIt \text{ or } zQ$$

Example 13.1   Given that the electrochemical equivalent of silver is 0.001118 gm per coulomb, what mass of silver will be deposited on the anode of a silver plating bath if a current of 2.5 A flows for 10 hours?

$$M = zIt$$
$$= 0.001118 \times 2.5 \times 10 \times 3600 \ (t \text{ in seconds})$$
$$= 100.62 \text{ g}$$

Electrolysis of copper sulphate solution

Figure 13.3 shows a suitable apparatus for the determination of the e.c.e. of copper. The figure is self-explanatory, but for the best results the following precautions should be taken. A suitable strength of electrolyte is 10 to 15 per cent solution by weight to which 5 cm³ of strong sulphuric acid is added to each 1000 cm³ of electrolyte. A suitable temperature is about 40° C. Both electrodes should be of pure copper sheet, the cathode being placed between two anodes so that copper can be deposited on both faces. The area of cathode, reckoning both sides, should be such that there is not less than 50 cm² per ampere. If the current density at the cathode surface is too high the copper deposit may be very rough, and particles may even fall off when the cathode is being washed. In any case this should be done very carefully, preferably with distilled water, and the cathode then dried in a current of warm air.

The experiment can be performed to determine the value of z for

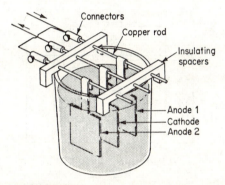

Figure 13.3   Arrangement of the copper sulphate electrolytic cell.

copper. Alternatively, given z, it can be used to check the accuracy of a current-measuring instrument such as an ammeter. The following values refer to an actual experiment:

Reading of ammeter	0.95 A
Mass of copper deposited	1.08 g
Time	1 h

In a copper sulphate solution the copper is divalent and z is equal to 0.000329 g/C

$$\therefore \; I = m/zt = 1.08/(0.000329 \times 3600)$$
$$= 0.912 \text{ A}$$

The ammeter therefore reads $0.95 - 0.912 = 0.038$ A too high.

The best results are obtained if silver nitrate, $AgNO_3$ is used for the electrolyte and pure silver plates used for anode and cathode.

The method is so accurate that it was used formerly as the basis for the definition of the ampere.

Primary cells

If a copper plate is immersed in water, copper ions leave the plate for the solution, the process therefore being as follows

$$Cu \rightarrow Cu^{2+} + 2e^-$$

The electrons are left behind in the plate which, as a result, acquires a negative potential with respect to the liquid. The value of this

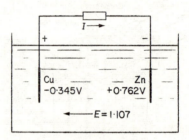

Figure 13.4   Simple cell.

potential is -0.345 V. Suppose a zinc plate is immersed, then the reverse process takes place, the zinc acquiring a positive potential, relative to the liquid of $+0.762$. The p.d. between zinc and copper is therefore $0.762 - (-0.345) = 1.107$ V and acts in the liquid from zinc to copper (figure 13.4). With respect to an external circuit the copper is therefore the positive plate and the zinc the negative, the e.m.f. of the cell being 1.107 V.

Actually no current could be taken from such a cell because of the very high resistance of water. This can be remedied by using dilute sulphuric acid and we then have what is called a simple cell. Suppose that such a cell is delivering current, then the dissociated ions of the acid, namely SO_4^{2-} and two H_3O^+ per molecule will move in directions the reverse of those of figure 13.2 because this motion is now the result of an internally produced e.m.f. and is not imposed from an outside source. From figure 13.5 the end results of the phenomena at the electrodes are now: at the anode the H_3O^+ ion has its charge neutralised by an electron coming from the outside circuit, after which

$$2H_3O = 2H_2O + H_2$$

Thus water is formed and hydrogen liberated. Two atoms of hydrogen combine to form hydrogen gas. At the cathode the SO_4^{2-} ion releases its charge, and two electrons enter the external circuit. The SO_4 without its charge immediately attacks the zinc cathode with the formation of zinc sulphate $ZnSO_4$, which dissolves. Thus no gas is given off at the zinc, but the plate is gradually consumed; in fact it is the energy set free by this chemical consumption of the zinc which

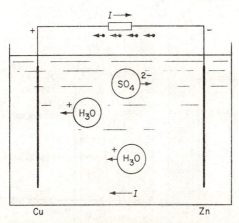

Figure 13.5   Ion movement in simple cell.

is converted into the electrical energy of the cell. Thus the zinc can be regarded as the 'fuel'.

The hydrogen given off at the anode is in the form of fine bubbles of gas which tend to adhere to the copper surface, thus forming a gas film which prevents further H_3O^+ ions reaching the copper, and which, in consequence, form a positive 'space charge' which repels the approaching ions. Thus, when delivering current, the p.d. of the cell falls off very rapidly to a low value and the current finally becomes zero. This effect is called polarisation.

Again, while the change to $ZnSO_4$ is an essential feature of the cell when delivering current, zinc of commercial variety is attacked by the acid irrespective of any corresponding production of available energy. This is because impurities in commercial zinc, such as iron, form minute cells with the zinc which are always in action, even when no current is being delivered to the external circuit. This is called local action. A successful primary cell must be free from both polarisation and local action.

The Leclanché Cell

This is the most widely used primary cell, and the depolarising is again carried out chemically. The agent is manganese dioxide,

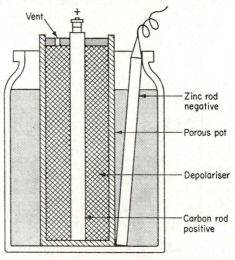

Figure 13.6   Construction of the Leclanché cell.

MnO_2, which is rich in oxygen but parts with some of this oxygen fairly readily. In the presence of atomic hydrogen and water it is able to oxidise the hydrogen to form additional water. The usual form of cell consists of a glass jar containing a solution of ammonium chloride, NH_4Cl. Standing in this is a zinc rod and a porous pot containing a carbon rod or plate as anode. Round this rod is packed a mixture of crushed carbon and manganese dioxide (figure 13.6). As with other cells zinc atoms migrate as ions into the solution thereby producing a surplus of electrons in the zinc itself.

$$Zn \rightarrow Zn^{2+} + 2e^-$$

On completing the circuit these electrons can migrate to the external circuit to form the external current and the following changes then take place in the cell

$$Zn + 2NH_4Cl = ZnCl_2 + 2NH_3 + H_2$$

It is the conversion of zinc to zinc chloride which provides the energy of the cell. The zinc chloride, $ZnCl_2$ and the ammonia, NH_3, are both water soluble. The H_3O^+ ions travel to the carbon rod and atomic hydrogen is liberated. This attacks the MnO_2 which oxidises the hydrogen to water thus

$$2H + 2MnO_2 = Mn_2O_3 + H_2O$$

As the manganese dioxide cannot act quickly on the liberated hydrogen the e.m.f. falls off considerably if the cell is made to deliver current over a long period. The best application for the cell is thus one requiring intermittent operation, as in bell circuits. The e.m.f. is about 1.5 V and the internal resistance 1 to 2 Ω. The great advantage of the cell is that it requires no attention beyond make up of the solution by adding water if evaporation takes place, and its very infrequent replacement by fresh solution.

An improved form of cell is the 'sack' cell in which the porous pot is replaced by a wrapping of textile fabric held in place by twine. The smaller space occupied by the fabric enables a greater amount of $MnO_2/$ carbon mixture to be used thus improving the depolarising properties and lowering the resistance. The zinc is in the form of a cylinder instead of the usual rod. This construction gives an internal resistance as low as 0.075 Ω for large cells of about 15 cm diameter and 30 cm height.

The dry cell is a modification of the Leclanché cell. First the glass or earthenware jar is eliminated by making the zinc in the form of a canister. Secondly, the liquid electrolyte is replaced by a moist paste consisting of a mixture of plaster of paris, flour, ammonium chloride and zinc chloride. The latter substance, being highly hygroscopic, is added to maintain the moistness of the paste. The carbon rod is surrounded by the usual manganese dioxide-carbon mixture, which is enclosed in a fabric sack. A section of a typical cell is shown in figure 13.7. The top of the cell has a layer of sawdust which acts as a bed for a sealing layer of bitumen. Any gases which form escape by vents, not shown, through the sealing layer. As the cell is similar to the sack cell the resistance is very low. Thus a new cell of 7 cm diameter and 18 cm high has an e.m.f. of 1.527 V and resistance

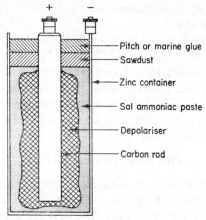

Figure 13.7   Construction of a dry cell.

0.18 Ω. The keeping properties of such large cells are good, but very small cells deteriorate rapidly and it is unwise to store them for any length of time.

The Weston Standard Cell

The function of the cells previously described is to provide current for some consuming device; for example, the Leclanché cell is commonly used to actuate electric bells. The standard cell is not used in this way; if it delivers more than a minute current it may be damaged. It is used solely as a voltage reference, as explained in chapter 19. This is because its e.m.f. is known to a high order of accuracy if the cell is made according to specification.

Its construction is illustrated in figure 13.8. The container is a H-shaped glass vessel with electrodes sealed into the bases of the two limbs. The positive electrode is pure mercury covered with a layer of a depolarising paste of cadmium sulphate mixed with mercurous sulphate. The negative electrode is an amalgam of cadmium with mercury. The electrolyte is a solution of cadmium sulphate, and in the 'normal' cell this solution is maintained in the saturated state by means of cadmium sulphate crystals placed immediately above the electrodes and below the restrictions in the two limbs. These crystals become lightly cemented together in the form of plugs. Either pure water or dilute sulphuric acid can be used as the solvent for the cadmium sulphate in the electrolyte, the latter being generally used because the properties of the cell are more permanent. From the figure it will be seen that the cell is hermetically sealed, and no organic substance is used. The e.m.f. of the cell is 1.0186 V at 20° C. The temperature coefficients of the two limbs are,

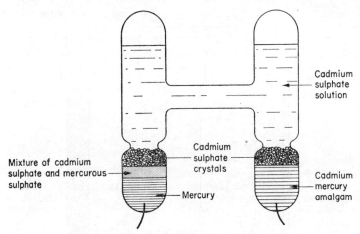

Figure 13.8   The Weston standard cell.

individually, somewhat high, but as one is positive and the other negative the total coefficient is small, -40μ V per °C between 15° C and 25° C. To keep the temperatures in the two limbs equal the cell is very commonly immersed in oil.

The Mercury Cell

For some purposes, for example, hearing aids and the supply to electrically driven watches in which the orthodox escapement is replaced by a small electrically driven tuning fork, very tiny cells are required. Also the ratio of the output to the mass must be very high in relation to other forms of cell, and the e.m.f. must remain reasonably constant over long periods. The mercury cell fulfils these requirements. Its construction is shown diagrammatically in figure 13.9. In plan the cell is circular, the construction having radial symmetry. In the middle is a hollow zinc cylinder, which is the negative electrode. Surrounding this is the electrolyte, which is a

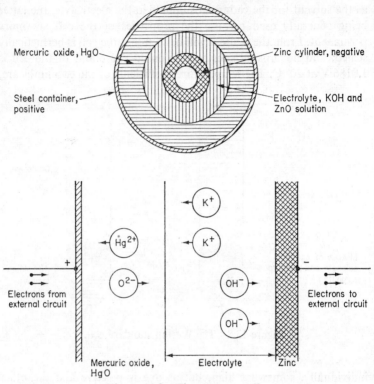

Figure 13.9   The mercury cell.

concentrated aqueous solution of KOH and ZnO. This is surrounded by the positive electrode, which is mercuric oxide HgO, mixed with very finely divided graphite, this having no effect on the action, being added to give the cell a low internal resistance. The whole is housed in a steel container which makes contact with the positive electrode. The lid of this case makes contact with the negative electrode and the seal between lid and container acts as the insulator between the two; thus the container is the positive terminal and the lid the negative terminal.

When the cell is delivering current the migration of the ions is as shown in the figure. In the electrolyte the K^+ ions move towards the HgO positive electrode while, in the electrode itself Hg^{2+} ions move towards the steel casing (the positive), while O^{2-} ions move towards the electrolyte. At the junction two K^+ ions are neutralised by one

Hg^{2+} ion and at the positive the charges on each Hg^{2+} ion are neutralised by the acceptance of two electrons from the external circuit.

At the negative electrode two OH^- ions give up their electrons to the zinc and thus to the external circuit, thereby maintaining a balance between the electrons entering the external circuit at the negative terminal and those leaving the external circuit at the positive terminal. The OH^- ions, after giving up their electrons combine with the zinc to form zinc oxide and water.

As no gases are evolved there is no polarisation and, in consequence, the cell maintains its e.m.f. of about 1.3 V for long periods.

Accumulators or Secondary Cells

A primary cell is one in which chemical energy is converted into electrical energy. A secondary cell is one in which electrical energy is accepted for the purpose of effecting a chemical change of a reversible nature, electrical energy being subsequently taken from the cell. Thus it is not electrical energy which is stored, or accumulated, but chemical energy.

The principle involved can be investigated by means of the apparatus shown in figure 13.10. A vessel containing dilute sulphuric acid has two lead electrodes which can be supplied with current from a battery B when the switch S is on contact 1. On contact 2 the cell is placed in the circuit of the ammeter A and resistor R. First of all the switch is placed on contact 1 so that current flows through the cell from anode to cathode. The cathode retains its metallic appearance but the anode becomes coated with a dark brown layer of lead

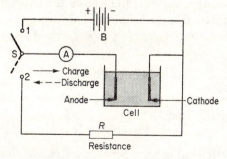

Figure 13.10 Experimental illustration of the action of a lead-acid accumulator.

peroxide PbO_2. The ions are H_3O^+ and SO_4^{2-} as in figure 13.2, and the liberation of hydrogen gas at the cathode takes place. The SO_4^{2-} ions on reaching the anode become uncharged SO_4 radicals and as such are compelled to take part in some chemical change. Some of them attack the lead of the anode to form lead sulphate $PbSO_4$ while others take oxygen from the water to become sulphuric acid, the reaction being

$$H_2O + SO_4 = H_2SO_4 + O$$

The oxygen is in atomic form and, as it is liberated in contact with the lead it combines with the lead to form dark brown lead peroxide PbO_2.

If the switch is now changed over to position 2 it will be found that the cell is now a source of electrical energy, the direction of its current output during 'discharging' being the reverse of that while

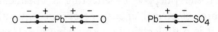

Figure 13.11   Molecular structure of lead oxide and lead sulphate.

'charging'. This current will be of short duration, but if the cycle of charging and discharging is repeated for some considerable time the current will flow for a much longer time. This is because the repeated chemical changes will penetrate deeper and deeper in the anode thereby giving a greater thickness of PbO_2, the substance in which the chemical energy is stored. On discharge there is conversion of PbO_2 to $PbSO_4$. In the lead oxide the lead is tetravalent but in the sulphate it is divalent and this explains the conversion of chemical to electrical energy. Figure 13.11 shows the structure of the two molecules. In the oxide the lead can be written

$$Pb^{4+} \text{ or } Pb - 4e$$

In the sulphate it is

$$Pb^{2+} \text{ or } Pb - 2e$$

At the same time, at the negative electrode which, when fully charged is metallic lead

$$Pb \rightarrow Pb^{2+} + 2e$$

showing the electrical balance between the two electrodes.

Because of the formation of lead sulphate the density of the acid falls during discharge but it rises again during charge. The usual value is 1.21 when fully charged, corresponding to 28 per cent by weight of acid. When discharged to 1.8 V the density is 1.18. The resistivity of the acid varies with the density being a minimum at 60° F. The e.m.f. when fully charged is 2.2 V and should not be allowed to fall below 1.8 V otherwise there will be excessive formation of lead sulphate.

Practical Accumulator Plates

The process of making accumulator plates by the above experimental process, called 'forming', is much too lengthy and expensive for commercial purposes. The practical method is to use plates in the form of cast grids, the holes of which are filled with a paste of red lead, Pb_3O_4 for the positive and litharge, PbO, for the negative. These are immersed in dilute sulphuric acid and the cell charged by passing current through it. This oxidises the Pb_3O_4 at the positive to the required PbO_2 and reduces the PbO at the negative to metallic lead. Even this process is slow for present-day economics, and it is therefore usual to add to the electrolyte certain chemicals such as nitrates, chlorates and acetates, which accelerate the reactions.

Because the energy exchanges take place at the positive plate this has to be of more robust construction than the negative. For all types the negative is of the cast grid type but for high discharge rates special constructions are necessary for the positive. Figure 13.12 shows a few examples. Figure 13.12a shows the ordinary grid used for all negatives and low discharge-rate positives. Figure 13.12b shows the 'chloride' positive plate and figure 13.12c the Exide positive plate used for very heavy duty, such as batteries for battery-driven locomotives.

Efficiency

This can be reckoned in two ways:

$$\text{Ampere hour or quantity efficiency} = \frac{\text{ampere hours of discharge}}{\text{ampere hours of charge}}$$

$$\text{Watt hour or energy efficiency} = \frac{\text{watt hours of discharge}}{\text{watt hours of charge}}$$

$$\text{or} \quad \frac{\text{joules of discharge}}{\text{joules of charge}}$$

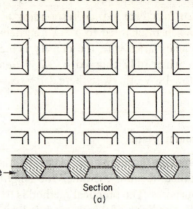

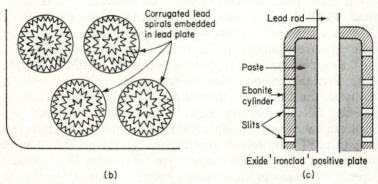

Figure 13.12   Types of accumulator plate.

If V is the terminal p.d. of a cell, E_1 the e.m.f. on charge and E_2 the e.m.f. on discharge, then

$$V_1 = E_1 + RI$$
$$\text{and} \quad V_2 = E_2 - RI$$

The average terminal p.d. during charge is therefore greater than during discharge, the difference being

$$V_1 - V_2 = (E_1 - E_2) + 2RI$$

The e.m.f. is greater during charging than during discharging, and it follows that the energy efficiency is less than the quantity efficiency. For a large cell representative values are 90 per cent for quantity and 75 per cent for energy efficiency. These are for the normal operation of a reduction of the e.m.f. to 1.8 V in 10 hours. This is called the 10-hour rating. If the discharge current is increased so as to reduce the time taken to fall to 1.8 V then both efficiencies are reduced. Conversely a reduced current increases both efficiencies.

Internal Resistance

If the open-current e.m.f. E of a cell is measured and then the terminal p.d. V, while delivering a measured current I, the resistance is given by

$$R=(E-V)/I$$

This method can be used for primary cells because their internal resistance is high but when the resistance is very low as with secondary cells, the difference between E and V is very small, there being very little difference between the voltmeter readings. The slightest error in reading may therefore lead to large errors in the calculated result. Methods involving small differences of large quantities should therefore be avoided whenever possible.

Figure 13.13 shows two satisfactory methods. Figure 13.13a shows the cell A test under test put 'back to back' with another cell B which has been under discharge for sufficiently long to give it a steady e.m.f. The voltmeter V can be a low-reading instrument, say 0–0.5 V, but care must be taken to connect it between the two positive terminals, as shown. With S open, that is the cell A in open circuit, the voltmeter reading is taken. This reading should be deducted, or added, to subsequent readings, depending on its sign. The resistance R is adjusted to give a high discharge rate, say 1-hour, and S closed. The voltmeter reading, corrected as above, will now

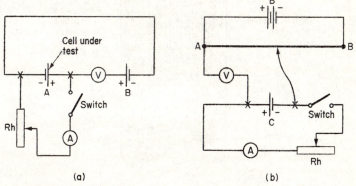

(a)   (b)

Figure 13.13   Two methods of determining the internal resistance of an accumulator cell.

give a reading roughly equal to the internal drop. The switch is now opened, and it will be observed that there is a sudden drop in the voltmeter reading, followed by a further gradual drop. It is the sudden, or momentary, change in voltage which is required, and this, divided by the current, will give the cell resistance.

Figure 13.13b shows an alternative method; the e.m.f. of the cell C now being balanced by the volt-drop along a stretched high-resistance wire AB. The battery B can be two Leclanché cells or three Daniell cells as it need supply current only along the wire, and there must be sufficient cells to give a volt-drop along AB greater than the cell e.m.f. The procedure is then as for method (a).

The Alkaline Accumulator

The positive electrode is a perforated steel tube, or containers containing nickel hydroxide. The negative electrodes are also perforated steel containers packed with finely divided iron, or, in another type, a mixture of iron and cadmium. The construction is illustrated in figure 13.14. The electrolyte is an aqueous solution of

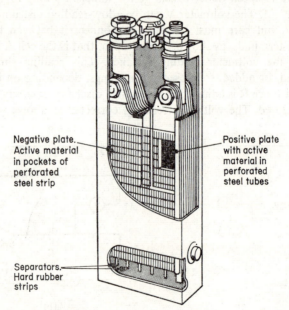

Negative plate.
Active material
in pockets of
perforated
steel strip

Positive plate
with active
material in
perforated
steel tubes

Separators.
Hard rubber
strips

Figure 13.14 Sectional view of typical alkaline accumulator. (*Courtesy of Alkaline Batteries Ltd.*)

potasium hydroxide. There is no change in the electrolyte during discharge or charge and the current can be much higher than in a lead-acid cell of comparable size. The e.m.f. per cell is 1.2 V. The efficiencies are lower than for lead-acid cells, representative values being 80 per cent for quantity and 60 per cent for energy efficiency.

Accumulator calculations

Let a battery of accumulator cells consist of n cells in series, each of e.m.f. E and internal resistance B. Let this battery be in series with a resistor R and let it be charged from a supply of p.d.; then, charging current

$$I = \frac{V - nE}{R + nB}$$

This shows that I can be varied (a) by varying V, that is charging is done by a separate generator, or (b) by varying R.

Example 13.2.   A battery of 50 cells each of e.m.f. 1.8 V and resistance 0.05 Ω is charged from a generator giving a terminal voltage of 120 V. If the conducting leads are of 0.25 Ω resistance and there is no other resistance in circuit, what will be the charging current (a) at the commencement, (b) when the e.m.f. per cell has risen to 2.2 V?

$$\text{(a)}\quad V = 120 \text{ V}, \quad nE = 50 \times 1.8 = 90 \text{ V}$$
$$nB = 50 \times 0.05 = 2.5 \ \Omega$$
$$\therefore \quad I = \frac{120 - 90}{0.25 + 2.5} = 10.9 \text{ A}$$
$$\text{(b) } nE = 50 \times 2.2 = 110 \text{ V}$$
$$\therefore \quad I = \frac{120 - 110}{0.25 + 2.5} = 3.64 \text{ A}$$

Example 13.3.   What will be the p.d. per cell in the two cases above?

$$\text{(a) p.d. per cell} = E + BI$$
$$= 1.8 + 0.05 \times 10.9$$
$$= 2.35 \text{ V}$$
$$\text{(b) } E + BI \qquad = 2.2 + 0.05 \times 3.64$$
$$= 2.38 \text{ V}$$

Example 13.4.   If the above battery is being charged from a 200 V supply and it is desired to keep the charging current at 20 A by including additional resistance in the circuit, what must be the value of this resistance (a) at the beginning, (b) at the end?

$$\text{(a)} \quad V = 200 \text{ V}, \ nE = 90 \text{ V}$$
$$R = nB + 0.25 + x = 2.75 + x$$
$$\therefore \quad 20 = \frac{200 - 90}{2.75 + x}.$$
$$20x + 55 = 110$$
$$x = 55/20 = 2.75 \ \Omega$$

$$\text{(b)} \ nE = 110 \text{ V}$$
$$\therefore \quad 20 = \frac{200 - 110}{2.75 + x}$$
$$20x + 55 = 90$$
$$x = 35/20 = 1.75 \ \Omega$$

14 ALTERNATING CURRENTS

Sinusoidal Quantities

Figure 14.1 shows a line OP which rotates with uniform angular velocity ω about the end O. If time is reckoned from the instant the line crosses the OX axis then the angle θ is given by

$$\theta = \omega t \text{ radians}$$

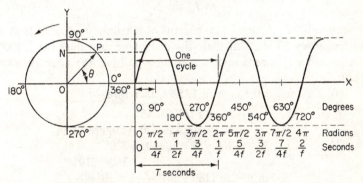

Figure 14.1   Generation of a sinusoidal wave.

The diagram also shows angles in degrees. For any position the projection ON on the OY axis is of length

$$ON = OP \sin \theta$$

Consequently if the length of ON is plotted against θ as in the right-hand diagram a wave-shaped curve corresponding to the sine law will be obtained. A quantity which varies with time according to such a law is called a sinusoidal quantity. The voltage and currents we shall consider are mainly of this type.

Suppose then that OP represents E_m, the maximum value of a sinusoidal e.m.f., then at any instant corresponding to the angle θ, the instantaneous value of the e.m.f. is

$$e = E_m \sin \theta = E_m \sin \omega t$$

The time taken for the rotating line to make one complete revolution is called the periodic time T. But one whole revolution is equal to 2π radians

$$\therefore \quad 2\pi = \omega T$$
$$T = 2\pi/\omega$$

The number of revolutions per second is the frequency f

$$\therefore \quad f = 1/T = \omega/2\pi$$
$$\therefore \quad \omega = 2\pi f$$

We can therefore write for the instantaneous value of a sinusoidal e.m.f.

$$e = E_m \sin 2\pi f t$$

and similarly for a sinusoidal current

$$i = I_m \sin 2\pi f t$$

Example 14.1. A sinusoidal current of maximum value 28.28 A is of frequency 50 Hz. Calculate the instantaneous values (a) 1/400th second after it is zero and is becoming positive, (b) after 19/600th second.

360° are turned through in 1/50 s, therefore

(a) $\qquad \theta = 360 \times 50/400 = 45°$
$\qquad\qquad i = 28.28 \sin 45° = 20$ A, figure 14.2a

(b) $\qquad \theta = 360 \times 50 \times 19/600 = 570°$
$$= 360 + 210°$$
$\therefore \sin \theta = \sin (180 + 30)° = -\sin 30° = -1/2$
$\therefore \quad i = -28.28 \times 1/2 = -14.14$ A. Figure 14.2b

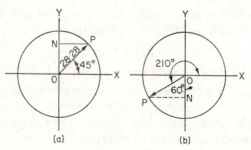

Figure 14.2

Root Mean Square (r.m.s.) Value

If all current and voltage waves were sinusoidal the magnitude could be expressed by the maximum value. Although the sinusoidal form is the ideal for most purposes there are many cases in which there is considerable departure from this form. We therefore require a method of expressing the magnitude which will apply to all waveforms, whatever their shape. Since the negative half wave is a repetition of the positive half with the sign changed it is sufficient to consider the half wave.

The basis of the method is the heating effect on a circuit with resistance only. Let an alternating current of constant maximum value and constant waveform pass through a given resistance for a definite time, then a certain amount of heat will be produced. Let a steady unidirectional current pass through the same resistance for the same time and let the value of this current be such that exactly the same amount of heat is produced. Then the magnitude of the alternating current is defined as being the same as that of the uni-directional current. Figure 14.3 shows a half wave, of any shape whatever. Let its duration be T', and let it be divided into a large

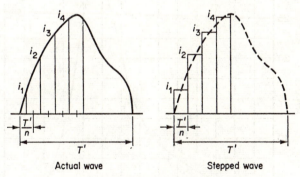

Actual wave   Stepped wave

Figure 14.3   Method of calculating r.m.s. value.

number, n, of equal intervals of T'/n. Let a mid-ordinate be erected at the middle of each interval and let the successive values be i_1, i_2, i_3, etc. Now let the actual wave be replaced by the stepped wave; provided we make the intervals very small the error due to using the

B.E.—6*

stepped, instead of the actual, wave is very small. Then, if the current flows through a resistor of R ohms

Heat generated in 1st interval $= 0.24\ i_1^2 R \times T'/n$ calories
Heat generated in 2nd interval $= 0.24\ i_2^2 R \times T'/n$ calories

and so on. Hence

Heat generated during time $T' = 0.24\ RT'\ (i_1^2 + i_2^2 + i_3^2 + ...)/n$

Now let I be the steady current which generates the same heat in the same time T', then

$$0.24\ I^2 RT' = 0.24\ RT'\ (i_1^2 + i_2^2 + i_3^2 + ...)/n$$
$$I = \sqrt{(i_1^2 + i_2^2 + i_3^2 ...)/n}$$

Example 14.2. The instantaneous values of an alternating current when measured at 15° intervals are -2, 1, 3.5, 5, 6.5, 8, 10.3, 15, 24.5, 30, 19, 7.3, 2, -1 amperes. Plot the half-wave of current and determine its average and r.m.s. values.

The curve is plotted in figure 14.4, and as it is a smooth curve, intervals of 15° giving 12 steps to the halfwave will give the answer

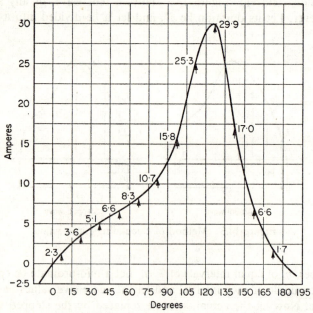

Figure 14.4   Calculation of r.m.s. value.

with very little error. The origin is chosen at a point where the curve crosses the OX axis. The mid-ordinates are marked by small arrows and their values indicated. The calculation is made in tabular form.

Mid-ordinate no.	Mid-ordinate height	(Mid-ordinate)²
1	2.3	5.29
2	3.6	12.96
3	5.1	26.01
4	6.6	43.56
5	8.3	68.89
6	10.7	114.49
7	15.8	249.64
8	25.3	640.09
9	29.9	894.01
10	17.0	289.00
11	6.6	43.59
12	1.7	2.89
	Sum = 132.9	Sum = 2390.42
	Mean = 11.08	Mean = 199.2
		Sq. root = 14.11

$\therefore$ r.m.s. value = 14.11 A and average value = 11.08 A

Example 14.3. A periodic current remains steady at 5 A for 0.01 s; rises suddenly to 10 A and remains steady for another 0.01 s. This cycle is repeated. Calculate the average and r.m.s. values of the current.

From figure 14.5a, since the intervals are of equal duration

$$I_{av} = (5+10)/2 = 7.5 \text{ A}$$

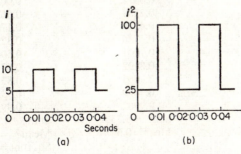

Figure 14.5

From figure 14.5b

$$(I^2)_{av} = (25 + 100)/2 = 62.5$$
$$\therefore \quad I = \sqrt{(62.5)} = 7.9 \text{ A}$$

The average and r.m.s. values for a sinusoidal wave are of particular importance. Taking a maximum value of 1.0 we can read the ordinates direct from tables. Also since the curve is symmetrical

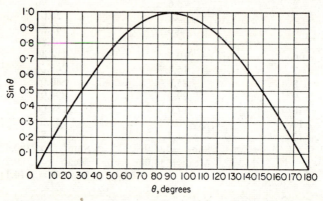

Figure 14.6   Plot of sin θ against θ for values of θ up to 180°.

about the 90° ordinate it is sufficient to take one quarter wave, that is from 0° to 90°. Ten-degree intervals are suitable for such a smooth curve. The half wave is given in figure 14.6 and the tabulation is below:

Mid-ordinate at 0°	Sin θ	Sin $^2\theta$
5	0.0872	0.0076
15	0.2588	0.0670
25	0.4226	0.1786
35	0.5736	0.3290
45	0.7071	0.5000
55	0.8192	0.6711
65	0.9063	0.8214
75	0.9659	0.9330
85	0.9962	0.9924
	Sum = 5.7389	Sum 4.5001
	Mean = 0.6374	Mean = 0.500
		Sq. root = 0.7071

Now $0.6374 = 2/\pi$ and $0.7071 = 1/\sqrt{2}$
Hence for the instantaneous and r.m.s. values of sinusoidal currents and e.m.f.s we have, since $\theta = \omega t$

$$i = I_m \sin \omega t, \qquad e = E_m \sin \omega t$$
$$I = I_m/\sqrt{2} \qquad E = E_m/\sqrt{2}$$
$$= 0.707 I_m \qquad = 0.707 E_m$$
$$I_{av} = (2/\pi) I_m \qquad E_{av} = (2/\pi) E_m$$

The ratio of r.m.s. to average value is called the form factor. For a sinusoidal wave it is therefore $0.7071/0.6374 = 1.11$.

Example 14.4. A sinusoidal alternating current has a maximum value of 15 A. What are its average and r.m.s. values?

$$I_{av} = 0.637 \times 15 = \ 9.56 \text{ A}$$
$$I = 0.707 \times 15 = 10.61 \text{ A}$$

Phasor diagrams

We saw from figure 14.1 that a sinusoidal wave can be drawn by projecting from a rotating line whose length represents the maximum value of the wave. Consequently, for most purposes, this part of the figure is all that is necessary. Steps in a further simplification are given in figure 14.7. First the axes OX and OY, being of fixed direction can be removed unless they are particularly required; this leaves the rotating line representing the maximum value of the quantity concerned, an e.m.f. in the case figured. Finally, since the r.m.s. value E is a definite fraction of E_m, the line can represent this value. Also the arrow indicating rotation can be omitted since the counterclockwise direction is now universally adopted. This leaves the line of length proportional to E. It is commonly called a vector but a better name is phasor, because it is associated with revolution, and its angular position at any instant is its phase at that instant.

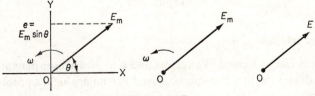

Figure 14.7

Phase difference

If there are two phasors, say one of e.m.f. and the other of current, the angle between them will be constant because both rotate at the same angular velocity of $\omega = 2\pi f$. This angle is called the phase difference. Figure 14.8 shows three cases, the wave diagrams as well as the phasor diagrams being drawn. In case 14.8a E and I are in phase. In case 14.8b I lags E and therefore E leads I. In case 14.8c I leads E which means that E lags I. If the phase difference is 90°, $\pi/2$ radian, or $1/4f$ s, the two phasors are said to be in quadrature.

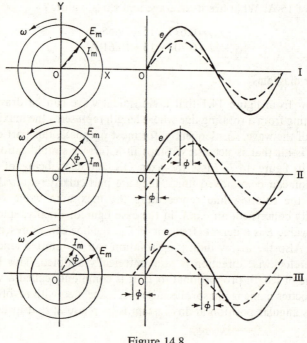

Figure 14.8

Phasor addition

Phasors can be treated like vectors and once the phasor diagram has been drawn the problem can be solved like an analogous problem in statics.

Example 14.5. An e.m.f. $E_1 = 80 \sin \omega t$ and an e.m.f. $E_2 = 40 \sin (\omega t - \pi/6)$ act together in a circuit. Find the resultant e.m.f. both in magnitude and phase.

Using maximum values the phasor diagram is that of figure 14.9. E_2 lags E_1 by $\pi/6$ radians $= 30°$. It is convenient to take the OX axis along the direction of E_1 and to project E_{2m} onto this. Then

Total OX component $= 80 + 40 \cos 30°$
$$= 80 + 40 \times 0.866 = 114.6$$
Total OY component $= 0 + 40 \sin 30° = 20$
$\therefore$ Resultant $\quad E_m = (114.6^2 + 20^2)^{\frac{1}{2}} = 116.2$ V
$$\tan \phi = -20/114.6 = -0.1754$$
$$\phi = -10° \text{ approx.} = 10 \times \pi/180 = 0.175 \text{ rad}$$

The equation for E with respect to E_1 is therefore

$$E = 116.2 \sin (\omega t - 0.175)$$

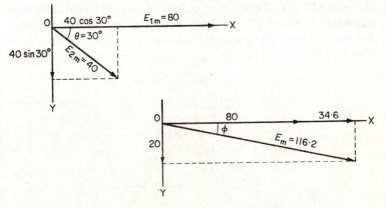

Figure 14.9 Phasor diagram for example 14.4.

Circuit with Resistance only

If a circuit possesses ohmic resistance only, it is not the seat of an induced e.m.f. and it does not store electricity. Hence all the applied p.d. has to do is supply the ohmic drop. Hence, at every instant

$$v = iR \text{ and } i = v/R$$

At every instant the ratio of instantaneous voltage v to instantaneous current i is the constant R, except that when $v=0$, $i=0$. Hence the waveform of current is identical with that of applied p.d., and the r.m.s. current I is in phase with the r.m.s. applied p.d., V

$$\therefore \ I=V/R$$

and if we put $v=V_m \sin \omega t$, then

$$i=I_m \sin \omega t$$

Again, since the energy expended is utilised solely in the production of heat, and the r.m.s. value is defined in terms of the heating effect, it follows that the power

$$P=VI=I^2R=V^2/R \text{ as in a d.c. circuit}$$

There is, however, a very important difference. The power varies from instant to instant since

$$P=vi=VI \sin {}^2\omega t$$
$$=\tfrac{1}{2}VI(1-\cos 2\omega t)$$

The term $\tfrac{1}{2}VI$ gives the power in watts which would be indicated by a wattmeter. The term $\tfrac{1}{2}VI \cos 2\omega t$ is an alternating quantity of double frequency $2f$, but whose average value over a whole period is zero. Figure 14.10 shows a voltage wave of maximum value $V_m=100$ and

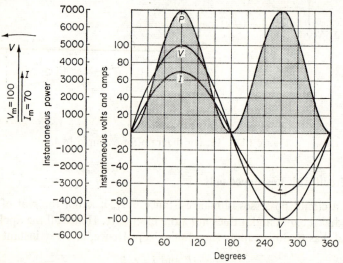

Figure 14.10 Voltage, current and power curves for a circuit having resistance only.

a current wave of $I_m=70$ in phase with it. The curve of power, cross-hatched, is obtained by plotting the products of instantaneous volts and amps. The maximum instantaneous power is $100 \times 70 = 7000$ watts. The average, which the reader can easily verify for himself, is 3500 watts.

Now $V=V_m/\sqrt{2}=70.7$ V and $I=I_m/\sqrt{2}=49.49$ A

Their product is

$P=70.7 \times 49.49 = 3500$ watts, showing that when V and I are in phase the average power is VI.

The shape of the power curve is explained by the fact that there is only a heating effect, that this is zero when $i=0$ and is a maximum when $i=I_m$. The instantaneous power is never negative.

Circuit with Self-inductance only

In this case the applied p.d. has only to overcome the self-induced e.m.f. But

$$\text{self-induced e.m.f.} = -L \times (\text{rate of change of current})$$

Hence, at any instant

$$v = +L \times (\text{rate of change of current})$$

Referring to figure 14.1 we see that an alternating current changes from $+I_m$ to $-I_m$ in $1/2f$ s.

$$\therefore \quad \text{Average rate of change} = 2I_m/(1/2f) = 4I_m f$$
$$\therefore \quad\quad\quad\quad V_{av} = 4LI_m/f$$
$$\text{Now } V_m = \pi/2 \times V_{av}$$
$$\therefore \quad\quad\quad\quad V_m = 2\pi f L I_m$$
$$\therefore \quad\quad\quad\quad V = 2\pi f L I = L\omega I$$
$$\therefore \quad\quad\quad\quad I = V/L\omega$$

The denominator $L\omega$ is called the inductive reactance X_L and the unit is obviously the ohm.

We must now find the phase difference between V and I. The current wave is drawn in figure 14.11. At any point, the rate of change is given by the slope of the tangent at that point. At the origin, where the current is zero it is obvious that the slope is a maximum. As we approach the crest of the wave the slope decreases and

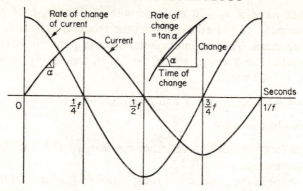

. Figure 14.11   Waves of current and rate of change of current.

becomes zero at the crest. Then the slope changes sign and progressively increases until the current is again zero, and so on. It follows that the curve of rate of change is another wave-shaped curve; it is, in fact, another sinusoid of the same frequency as the current but having its positive maximum at the current zero. Thus the current is in quadrature lagging with respect to the voltage. Hence if we put

$$v = V_m \sin \omega t$$
$$i = I_m \sin (\omega t - \pi/2)$$

Example 14.5.   An inductor whose coil is of negligible resistance has an inductance of 0.02H. It is connected to a supply at 100 V r.m.s., and frequency 50 Hz. Calculate the current

$$\omega = 2\pi f = 2\pi \times 50 = 314$$
$$X_L = L\omega = 0.02 \times 314 = 6.28 \ \Omega$$
$$\therefore \quad I = 100/6.28 = 15.91 \text{ A lagging } V \text{ by } 90°$$

The current can be written

$$I = 15.91 \underline{|90°}$$

In figure 14.12 curves of applied p.d. and of current are again drawn, but this time the current lags the voltage by 90°. We see that the power is again periodic of frequency 2f, but this time the average power is zero. Hence a wattmeter connected in such a circuit would register zero power, even with V and I finite. The explanation is the exchange of power between source and inductor as the magnetic

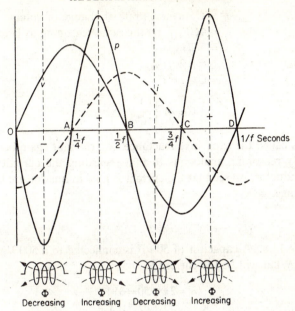

Figure 14.12 Variations in power during one cycle in a purely inductive circuit.

flux of the inductor increases and decreases. When the current is increasing the flux is increasing; the stored energy in it is increasing and the power is positive. When the current is decreasing the flux is decreasing, stored energy now being returned to the source. The power is therefore negative. Thus,

From O to A, current decreasing, flux decreasing, power negative.

From A to B, current increasing, flux increasing, power positive.

From B to C, as from O to A but with reversed current.

From C to D, as from A to B but with reversed current.

We see that the average power is zero.

Circuit with capacitance only

The fundamental relationship is

$$Q = CV$$

$$\therefore \text{ Current} \times \text{Time} = C \times \text{volts}$$

$$\text{Current} \qquad\quad = C \times (\text{volts/time})$$

$$\qquad\qquad\qquad = C \times \text{rate of change of voltage}$$

$$\therefore I_{av} = C \times \text{average rate of change of voltage}$$
$$= 4CV_m f, \text{ by the same reasoning as before}$$
$$\therefore 2/\pi I_m = 4CfV_m$$
$$I_m = C \times 2\pi f V_m$$
$$I = C \times 2\pi f V$$

$$= VC\omega \text{ or } \frac{V}{1/C\omega}$$

$1/C\omega$ is called the capacitive reactance, X_c, the unit again being the ohm. By reasoning analogous to that regarding the current taken by an inductor, we see that the current is now in quadrature leading the voltage.

Example 14.6. A capacitor of 50 μF is connected to a 500 V, 50 Hz supply. What will be the current?

$$\omega = 2\pi \times 50 = 314$$
$$C\omega = 50 \times 10^{-6} \times 314$$
$$\therefore I = 500 \times 50 \times 10^{-6} \times 314 = 7.85 \underline{|+90°} \text{ A}$$

The reactance is $X_c = 1/C\omega = 63.7 \ \Omega$

The only way in which a capacitor can accept energy is by an increase in its electric flux Φ. The magnitude of this flux is proportional to the p.d. and it increases and decreases as the p.d. increases and decreases. Hence there is an exchange of energy between source and capacitor, but in this case it follows the changes in p.d. These changes are shown in figure 14.13, and again the average power is zero. Consequently, whenever a current is in quadrature with a voltage the average power conveyed by the current is zero.

From O to A, p.d. increasing, electric flux increasing, power positive.
From A to B, p.d. decreasing, electric flux decreasing, power negative.
From B to C, p.d. increasing, electric flux increasing, power positive.
From C to D, p.d. decreasing, electric flux decreasing, power negative.

In all three cases, resistance only, inductance only, and capacitance only the power fluctuates at twice the supply frequency. We shall see that this is the case in all single phase circuits.

Figure 14.14 provides a useful recapitulation.

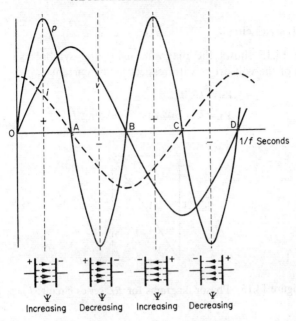

Figure 14.13 Variations in power during one cycle in a purely capacitive circuit.

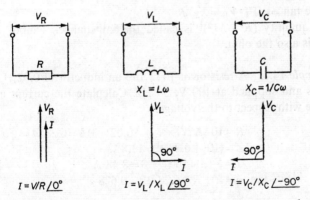

Figure 14.14 The behaviour of pure resistance, inductance and capacitance to an alternating applied p.d.

The RL series circuit

Figure 14.15 shows the diagram, using the symbols shown and phases of the volt-drops with respect to the current

$$V_{\mathrm{R}} = IR \underline{|0°}$$
$$V_{\mathrm{L}} = IX_{\mathrm{L}} \underline{|90°} \text{ that is leading } I$$

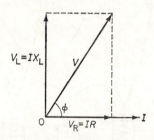

Figure 14.15 Phasor diagrams for *RL* series circuit: *I* lags *V*.

This gives the phasor diagram shown and

$$V^2 = V_{\mathrm{R}}^2 + V_{\mathrm{L}}^2$$
$$V = (V_{\mathrm{R}}^2 + V_{\mathrm{L}}^2)^{\frac{1}{2}} = I (R^2 + X_{\mathrm{L}}^2)^{\frac{1}{2}}$$
$$\therefore \quad I = \frac{V}{(R^2 + X_{\mathrm{L}}^2)^{\frac{1}{2}}} \underline{|\phi°} \text{ with respect to } V$$

where $\tan \phi = V_{\mathrm{L}} / V_{\mathrm{R}} = X_{\mathrm{L}} / R$

The quantity $(R^2 + X_{\mathrm{L}}^2)^{\frac{1}{2}}$ is called the impedance, symbol Z; its unit is also the ohm.

Example 14.7. A resistor of 10 Ω and an inductor of 0.02 H are in series and supplied at 100 V, 50 Hz. Calculate the current and its phase with respect to the voltage.

$$R = 10 \ \Omega, \ X_{\mathrm{L}} = L\omega = 0.02 \times 314 = 6.28 \ \Omega$$
$$\therefore \quad Z = (10^2 + 6.28^2)^{\frac{1}{2}} = 11.8 \ \Omega$$
$$\therefore \quad I = V/Z = 100/11.8 = 8.48 \ \mathrm{A}$$
$$\tan \phi = X_{\mathrm{L}}/R = 6.28/10 = 0.628$$
$$\therefore \quad \phi = 32°$$
$$\therefore \quad I = 8.48 \underline{|-32°} \ \mathrm{A}$$

Example 14.8. A coil of 5 Ω resistance is to be supplied from a source at 100 V, 50 Hz, but the current in it is to be 10 A. What inductance must be connected in series with it to achieve this?

$$Z = V/I$$
$$= 100/10 = 10\ \Omega$$

From $\quad Z^2 = R^2 + X_{\text{L}}^2$

$$X_{\text{L}}^2 = Z^2 - R^2 = 10^2 - 5^2 = 75$$

∴ $\quad X_{\text{L}} = 8.65\ \Omega$

∴ $\quad L = X_{\text{L}}/\omega = 8.65/314 = 0.0276\ \text{H}$

$\tan \phi \quad = X_{\text{L}}/R = 8.65/5 = 1.73$

$$\phi = 60°$$

∴ $\quad I = 10\ \underline{|-60°}\ \text{A}$

The RC series circuit

The phasor diagram is given in figure 14.16 and we have, with respect to I

$$V_{\text{R}} = IR\ \underline{|0°}$$

$$V_C = IX_C\ \underline{|90°}\ \text{that is lagging}\ I,\ IX_C\ \underline{|-90°}$$

∴ $\quad V^2 = V_{\text{R}}^2 + V_C^2 = I\ (R^2 + X_C^2)^{\frac{1}{2}}$

$$I = \frac{V}{(R^2 + X_C^2)^{\frac{1}{2}}}\ \underline{|-\phi}\ \text{with respect to}\ V$$

where $\tan \phi \quad = X_C/R$

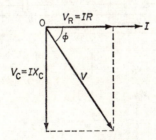

Figure 14.16   Phasor diagrams for *RC* series circuit: *I* leads *V*.

Example 14.9. A resistor of 2 Ω is in series with a capacitor of 600 μF. The applied p.d. is 100 V at 50 Hz. Calculate the current and its phase

$$\omega = 314 \text{ for 50 Hz}$$
$$\therefore X_C = 1/C\omega$$
$$= \frac{10^6}{600 \times 314} = 5.3 \ \Omega$$
$$\therefore Z = (2^2 + 5.3^2)^{\frac{1}{2}} = 5.67 \ \Omega$$
$$I = 100/5.67 = 17.63 \text{ A}$$
$$\tan \phi = X_C/R = -5.3/2 = -2.65$$
$$\phi = -69°20'$$
$$\therefore I = 17.63 \underline{|+69°20'} \text{ A}$$

The RLC series circuit

From the phasor diagram of figure 14.17 we have

$$V^2 = V_R{}^2 + (V_L - V_C)^2$$
$$= (IR)^2 + (IL\omega - I/C\omega)^2$$
$$= I^2 [R^2 + (L\omega - 1/C\omega)^2]$$
$$\therefore I = \frac{V}{[R^2 + (L\omega - 1/C\omega)^2]^{\frac{1}{2}}} \underline{|\phi°}$$
$$\text{and } \tan \phi = \frac{L\omega - 1/C\omega}{R}$$

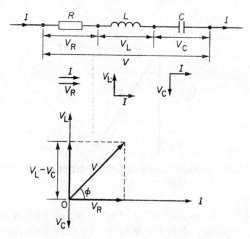

Figure 14.17   The *RLC* series circuit.

Example 14.10. A RLC series circuit has $R = 10 \ \Omega$, $L = 1.59$ mH, $C = 63.5 \mu$F. It is supplied at 20 V, 50 Hz. Find the current and its phase

$$R = 10 \ \Omega, \ L\omega = 0.5 \ \Omega, \ 1/C\omega = 50 \ \Omega$$
$$\therefore \ Z = [10^2 + (0.5 - 50)^2]^{\frac{1}{2}} = 50.5 \ \Omega$$
$$I = 20/50.5 = 0.396 \ A$$
$$\tan \phi = -49.5/10 = -4.95$$
$$\therefore \ \phi = 78° \ 35' \text{ leading the applied p.d.}$$
$$I = 0.396 \underline{|78°35'} \text{ (omitting the + sign)}$$

There are three possible cases of the *RLC* circuit. As shown in figure 14.18, $X_L > X_C$, $X_L = X_C$, $X_L < X_C$. The first of these is shown in figure 14.18i. The total reactance $X_L - X_C$ is positive and the current lags the applied p.d. In figure 14.18ii $X_L = X_C$. The total reactance is zero and therefore $Z = R$. The current is equal to V/R and is in phase with V. This is a very special case and it is examined below. In figure 14.18iii $X_L < X_C$. The capacitive reactance predominates and the current leads the applied p.d.

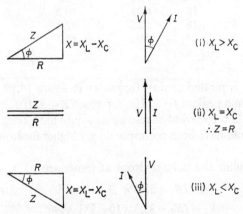

Figure 14.18   Effect of the relative values of X_L and X_C.

Resonance in the RLC series circuit

It is clear that if the frequency of the p.d. applied to a *RLC* circuit can be varied there will be one particular value which makes $L\omega$ equal to $1/C\omega$. The impedance will then be equal to R and the current will have the maximum possible value for a given value of V. When this occurs the phenomenon is called resonance. It is of great practical importance particularly in radio circuits, where the largest possible value of the current for a given applied p.d., is required. On the other hand if resonance occurs in power circuits it can have serious and undesirable consequences.

Consider a circuit in which $R=3\ \Omega$, $L=31.8$ mH and $C=318\ \mu$F. Let $V=100$ and let its frequency be varied from 10 Hz to 100 Hz. The effects of this frequency variation are conveniently shown in tabular form, thus

Frequency	R	$L\omega$	$1/C\omega$	$Z=[R^2+(L\omega-1/C\omega)^2]^{\frac{1}{2}}$	$I=V/Z$
10	3	2	50	48.1	2.1
20	3	4	25	21.2	4.7
30	3	6	16.67	11.1	9.0
40	3	8	12.5	5.4	18.5
50	3	10	10	3	33.3
60	3	12	8.33	4.7	21.3
70	3	14	7.14	7.5	13.3
80	3	16	6.25	10.2	9.8
90	3	18	5.55	12.8	7.8
100	3	20	5.0	15.3	6.5

The current is plotted against frequency in figure 14.19 and we see that it rises to a peak at $f=50$ Hz, for which $Z=R$. If R has been $1\ \Omega$ instead of $3\ \Omega$ the current at resonance would have been 100 A, and its value would have been correspondingly higher for lower values of R.

Now calculate the voltage drops at resonance

$$V_R=IR\ =33.3\times\ 3=100\ \text{V}\underline{|0°}$$
$$V_L=IX_L=33.3\times10=333\ \text{V}\underline{|90°}$$
$$V_C=IX_C=33.3\times10=333\ \text{V}\underline{|-90°}$$

the phase angles all being with respect to I.

If R had been $1\ \Omega$ instead of $3\ \Omega$, V_L and V_X would each have been 1000 V, and for smaller values of R, greater still. We therefore see that the effects of resonance are twofold; first, a very large current, and second, high volt-drops across both inductor and capacitor, which can be many times the applied p.d. if R is small.

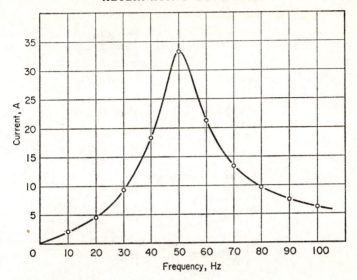

Figure 14.19   Resonance curve for *RLC* series circuit.

The explanation of this phenomenon is the exchange of energy, this time between the magnetic field of the inductor and the electric field of the capacitor.

The frequency of resonance or natural frequency f_0 is given by the relationship

$$L\omega = 1/C\omega$$

$$\therefore \quad \omega = \frac{1}{(LC)^{\frac{1}{2}}}$$

$$\therefore \quad f_0 = \frac{1}{2\pi(LC)^{\frac{1}{2}}}$$

Consider again the heavy truck as the mechanical analogue of inductance. Let it be fixed to a spring, as in figure 14.20, so that it can now only oscillate. The frequency of oscillations if the system is left to itself is determined by the product of the mass (inertia) and the extension or compression of the spring due to the application of unit force F (elasticity). For any velocity v the truck will have a kinetic energy of $\frac{1}{2} mv^2$. For any extension or compression d, the spring will have a potential energy of $\frac{1}{2}Fd$. $\frac{1}{2} mv^2$ is analogous to $\frac{1}{2}LI^2$ and $\frac{1}{2}Fd$ is analogous to $\frac{1}{2} VQ$.

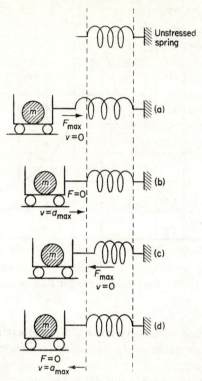

Figure 14.20   Oscillations in the mechanical analogue.

In figure 14.20a the truck is at the extent of its travel and the spring fully extended. The truck is momentarily at rest and so its kinetic energy, W_1, $\frac{1}{2} mv^2$, is zero. The spring is fully extended and therefore its potential energy, W_2, is a maximum. In figure 14.20b the truck has its maximum velocity and therefore W_1 is a maximum. The spring is in its unstressed position and its potential energy, W_2, is zero. Thus in regaining its unstressed position the spring has given up the whole of its energy to the truck. In figure 14.20c the velocity of the truck is again zero and the spring is fully compressed, the kinetic energy W_1 having been converted into potential energy, W_2. In figure 14.20d, v is a maximum, but in the reversed direction and the spring is unstressed so that W_2 has again been converted to W_1. The next stage is a repetition of process (a). We see that the fundamental property of the oscillating system is the exchange of energy between that part which provides the inertia and that which provides the elasticity.

Suppose that the truck is acted on by a series of impulses so timed that the direction of each is in the direction of motion, then the energy of these impulses will be added to W_1 which will progressively increase; consequently W_2 will increase at the same rate. This means an increase in the velocity attained by the truck and an increase in the extension of the spring, that is, of the amplitude of the oscillation. If the frequency of these timed impulses is the same as the natural frequency f_0 of the system then exceedingly large amplitudes will be produced.

This is what happens with the RL resonating system. The impressed frequency of the applied p.d. is equal to the natural frequency f_0.

If there is resistance this results in a loss of energy through conversion to heat energy in both the mechanical and the electrical systems and the greater the resistance the smaller the amplitude.

Branched Circuits

In a series circuit the total applied p.d. is the phasor sum of the separate voltage drops. In a branched circuit the total current supplied is the phasor sum of the separate branch currents. Consider the simple case of L and C in parallel, but no R, shown in figure 14.21.

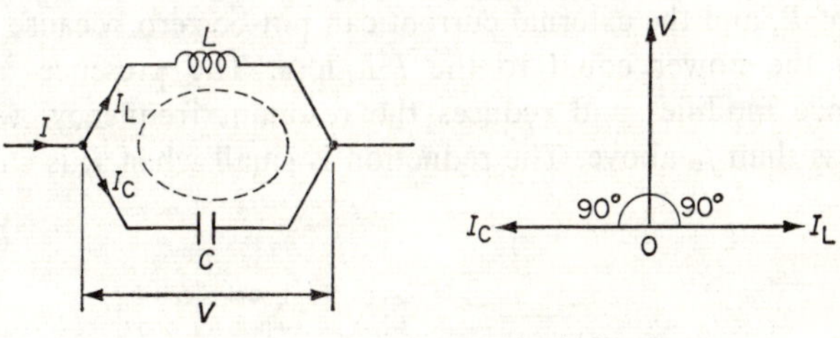

Figure 14.21 LC branched circuit.

$I_L = V/L\omega \underline{|-90°}$ and $I_C = VC\omega \underline{|+90°}$. Consequently the total current $I = I_L - I_C$ (phasor difference). If L and C are such that I_L and I_C are numerically equal we have the apparent paradox, that, although the current I fed to the circuit is zero, there is finite current in the two branches. The explanation is that L and C are in series relative to the dotted local circuit, so that we again have conditions corresponding to the analogue of figure 14.20. With zero resistance the oscillations will go on indefinitely because of the energy exchanges between the fields of the inductor and capacitor. Consequently the condition for resonance in a branched circuit is minimum current fed in from the outside, zero current in the theoretical case of zero resistance or other factors causing losses which result in heat production. Also there is no voltage rise, the p.d. across both inductor and capacitor being held at the impressed p.d. V. The condition for resonance in the LC branched circuit is

$$I_L = I_C$$
$$V/L\omega = VC\omega$$
$$\omega = (1/LC)^{\frac{1}{2}}$$
$$\text{and} \qquad f_0 = \frac{1}{2\pi(LC)^{\frac{1}{2}}}$$

as with the series circuit. We see that since $I = 0$, $Z = \infty$ to currents of frequency f_0, but not to other frequencies. This is the basis of rejector circuits which do not pass currents of frequency f_0, but allow currents of all other frequencies to pass.

Figure 14.22 shows a branched circuit in which one branch has ohmic resistance. There is again a local circuit, as shown dotted, with respect to which all three parameters are in series. Hence resonance can take place but the current will be controlled by the value of R, and the external current can not be zero because it must convey the power equal to the I^2R loss. The presence of finite resistance modifies and reduces the resonant frequency, which is now less than f_0 above. The reduction is small when R is small.

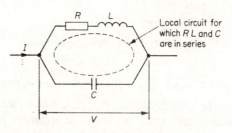

Figure 14.22   Branched circuit with resistance.

Example 14.11. A coil of resistance 5 Ω and inductance 31.8 mH is in parallel with a series circuit of resistance 2 Ω and capacitance 159 μF. If the supply is 200 V, 50 Hz, calculate the total current and the impedance of the whole.

$$X_1 = 314 \times 0.0318 = 10 \ \Omega$$
$$\therefore \ Z_1 = (5^2 + 10^2)^{\frac{1}{2}} = 11.18 \ \Omega$$
$$I_1 = 200/11.18 = 17.9 \ A$$
$$\phi_1 = \tan^{-1} 10/5 = 63°26' \text{ lagging}$$
$$\therefore \ I_1 = 17.9 \ \underline{|-63°26'} \ A$$
$$X_2 = 1/(314 \times 159 \times 10^{-6}) = 20 \ \Omega$$
$$\therefore \ Z_2 = (2^2 + 20^2)^{\frac{1}{2}} = 20.1 \ \Omega$$
$$I_2 = 200/20.1 = 9.95 \ A$$
$$\phi_2 = \tan^{-1} 20/2 = 84° \ 18' \text{ leading}$$
$$\therefore \ I_2 = 9.95 \ \underline{|84° \ 18'} \ A$$

Treating the phasors like two forces acting at a point we can use the cosine formula, giving

$$I^2 = 17.9^2 + 9.95^2 + 2 \times 17.9 \times 9.95 \cos (63°26' + 84° \ 18')$$
$$= 320 + 99 - 301 = 118$$
$$\therefore \ I = 10.9 \ A \text{ and } Z = 200/10.9 = 18.3 \ \Omega$$

Power Factor

Consider a current I lagging the p.d. V by an angle ϕ, as in figure 14.23. We can imagine the current resolved into two components at right-angles to one another.

     (i) the component $I \cos \phi$ in phase with V
    (ii) the component $I \sin \phi$ in quadrature with V.

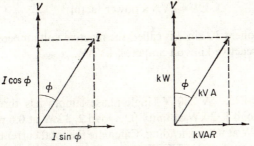

Figure 14.23   Resolution of a current into two components at right angles.

Now the power conveyed by a current in phase with a voltage is the product of that current and the voltage.

Hence power conveyed by the $I \cos \phi$ component $= VI \cos \phi$. The power conveyed by a current in quadrature with the voltage is zero, and therefore the $I \sin \phi$ component makes no contribution to the power

$$\therefore \quad W = VI \cos \phi$$

Example 14.12. *RL* series circuit with $R = 8.66 \ \Omega$, $L = 0.0159$ H. $V = 100$, $f = 50$ Hz

$$X_\mathrm{L} = 0.0159 \times 314 = 5 \ \Omega$$
$$\therefore \quad Z = (8.66^2 + 5^2)^{\frac{1}{2}} = 10 \ \Omega$$
$$\therefore \quad I = 100/10 = 10 \ \mathrm{A}$$
$$\tan \phi = 5/0.866 = 0.577$$
$$\phi = 30°, \cos \phi = 0.866$$
$$\therefore \quad P = VI \cos \phi = 100 \times 10 \times 0.866$$
$$= 866 \ \mathrm{W}$$

In what is called a 'passive circuit', that is, one in which there is no appliance such as an electric motor, and therefore one in which there is no expenditure of energy other than that required to generate the joule heat, the power is equal to $I^2 R$

$$\therefore \quad P = 10^2 \times 8.66 = 8.66 = 866 \ \mathrm{W}, \text{ as before}$$

Cos ϕ is called the power factor. The product VI is called the volt amperes. Hence

$$\text{Watts} = \text{volt amperes} \times \text{power factor}$$
$$\text{kW} = \text{kVA} \times \text{power factor}$$

The component $VI \sin \phi$ is called the reactive volt amperes and kVA $\sin \phi$ the reactive kilovolt amperes, kVAR.

Example 14.13.   A 240 V single-phase supply has three loads as follows: Load 1, 2 kW at unity p.f.; load 2, 3 kW at 0.6 p.f. lagging; load 3, 1 kW at 0.8 p.f. leading. Calculate the total current and power and the power factor of the supply.

Load 1
$$P_1 = VI_1 \cos \phi_1$$
$$2000 = 240\, I_1 \times 1.0$$
$$I_1 = 8.33 \text{ A}$$

Load 2
$$P_2 = VI_2 \cos \phi_2$$
$$3000 = 240\, I_2 \times 0.6$$
$$I_2 = 20.8 \text{ A}$$

Load 3
$$P_3 = VI_3 \cos \phi_3$$
$$1000 = 240\, I_3 \times 0.8$$
$$I_3 = 5.2 \text{ A}$$

The in-phase components are
8.33; $5.2 \times 0.8 = 4.16$, and $20.8 \times 0.6 = 12.46$, giving a total in-phase current of 24.95 A, figure 14.24.

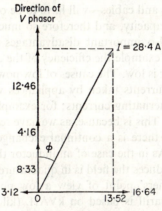

Figure 14.24

Since $\sin \phi_2 = 0.8$ and $\sin \phi_3 = 0.6$ the quadrature components are

$$20.8 \times 0.8 = 16.64 \text{ A lagging}$$
and
$$5.2 \times 0.6 = 3.12 \text{ A leading}$$

The total quadrature, or reactive component is therefore 13.52 A lagging. The total current is
$$I = (24.95^2 + 13.52^2)^{\frac{1}{2}} = 28.4 \text{ A}$$
The total kVA $= (240 \times 28.4)/1000 = 6.82$
The total kW $= 2 + 3 + 1 = 6$ kW
$\therefore$ Resultant p.f. $= 6/6.82 = 0.88$ lagging

B.E.—7

Power factor improvement by capacitor

In the above example, if the power factor of all three loads could have been unity the current would have been

$$I = 1000 \times kW/V = 6000/240 = 25.0 \text{ A instead of } 28.4 \text{ A}$$

Suppose that 20 000 kW are to be supplied at 20 000 V. At unity p.f. the current is

$$I = \frac{20\,000 \times 1000}{20\,000} = 1000 \text{ A}$$

If the power factor is 0.8 then the current, for the same real power, is now

$$1000/0.8 = 1250 \text{ A}$$

Thus everything which carries the current—the alternator, switch-gear, transformers and cables—will have to be of 25 per cent greater current-carrying capacity, and therefore so much larger and expensive. There are also important disadvantages from the operating point of view; for example the efficiency of the alternator is reduced if the power factor is low. The causes of low power factor are mainly the magnetising currents taken by appliances whose magnetic flux is produced by alternating currents; for example, transformers and induction motors. This is because, as we have seen, with an alternating magnetic flux there is a continual exchange of energy between field and source. As in the case of an inductor the component of the current which produces the field is in quadrature lagging the voltage. From the consumers' point of view a low power factor is a disadvantage if the tariff is based on kVAH, (kilovolt ampere hours) and not just on kWH.

When the conditions are appropriate the obviously simplest method is to connect a capacitor to the offending appliance. Figure 14.24 shows a current I_1 lagging the p.d. V by the angle ϕ. If a capacitor taking a current I_C is connected in parallel with the appliance, the total current will be I_2 in phase with V. Hence, although the appliance current remains at I_1, the current from the supply is reduced to I_2 and the power factor increased from $\cos \phi$ to unity.

Example 14.14.   A certain single-phase installation takes a load of 25 kW at 500 V, 50 Hz, the power factor being 0.8. What capacitance must be connected in parallel with the load to bring the overall power factor up to unity?

$$kW = 25$$
$$kVA = kW/\cos \phi = 25/0.8 = 31.25$$
$$kVAR = kVA \sin \phi = 31.25 \times 0.6 = 18.75 \text{ (figure 14.25)}$$
$$\therefore \quad kVAR_C = 18.75$$

But $\quad kVAR_C = VI_C/1000$

$$\therefore \quad 18.75 = 500I_C/1000 = I_C/2$$
$$I_C = 37.5 \text{ A}$$
$$\therefore \quad C = I_C/V\omega = 10^6 \times 37.5/(500 \times 314)$$
$$= 238.5 \mu F$$

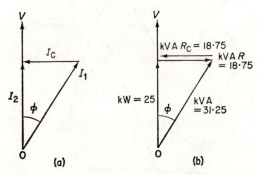

Figure 14.25

Complex Waveforms

Although the sinusoidal waveform is the ideal it frequently happens that voltage and current waves are not of this ideal shape. Examples are given in chapter 18. With a true alternating quantity having an average value of zero but of waveform different from the sinusoidal it can be shown that the actual wave is equivalent to the sum of a series of sinusoidal waves having frequencies of f, $3f$, $5f$, etc, where f is the supply frequency. The component of frequency f is called the fundamental and the others the harmonics. The fundamental is generally, but not always, greater in amplitude than any of the harmonics.

As an example consider a voltage wave for which the crest value of the fundamental is 100 V and the frequency 50 Hz. Let there be a third harmonic of crest value 50 V; its frequency will be 150 Hz. Assuming that fundamental and harmonic are zero and increasing in a positive sense at the same instant the resultant and component waves will be as shown in figure 14.26a.

(a) Let the p.d. be applied to a pure resistor of 1 Ω.

$$I_{1m} = V_{1m}/R = 100/1 = 100 \text{ A at } f = 50 \text{ Hz}$$
$$I_{3m} = V_{3m}/R = 50/1 = 50 \text{ A at } f = 150 \text{ Hz}$$

Thus the resultant current wave is of the same shape as the voltage wave.

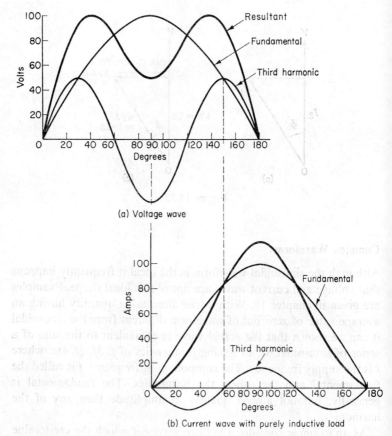

(a) Voltage wave

(b) Current wave with purely inductive load

Figure 14.26   Voltage and current waves with a purely inductive load.

(b) Let the p.d. be applied to a pure reactance of $X_L = 1\ \Omega$ at $f = 50$ Hz; then its reactance at 150 Hz will be $3\ \Omega$

$\therefore\ I_{1m} = 100/1 = 100$ A at $f = 50$ Hz lagging V_1 by $90°$
$I_{3m} = 50/3 = 16.7$ A at $f = 150$ Hz lagging V_3 by $90°$

where $90°$ now corresponds to one-quarter of a cycle of the 150 Hz wave.

The components and resultant wave are given in figure 14.26b and we see that the current wave is much nearer to the sinusoidal form than the wave of applied p.d. This smoothing effect of an inductor is of great importance in cases where the waveform departs considerably from the sinusoidal (see chapter 18).

(c) Let the p.d. be applied to a pure capacitor of reactance $X_C = 1\ \Omega$ at 50 Hz. Then its reactance at 150 Hz will be $1/3 = 0.33\ \Omega$

$I_{1m} = 100/1 = 100$ A at $f = 50$ Hz leading V_1 by $90°$
$I_{3m} = 50/0.33 = 150$ A at $f = 150$ Hz leading V_3 by $90°$

We see that, with a pure capacitor as load, the harmonics in the current wave are amplified in proportion to their frequencies. Thus, in the example the harmonic is of greater ampliture than the fundamental. The student should draw, as an exercise, the current wave for this case.

15 THREE-PHASE WORKING. ROTATING MAGNETIC FIELDS

The winding in which the e.m.f. is induced in a three-phase source consists essentially of three separate windings so disposed that there is a phase difference of 120° between each pair. Diagrammatically the arrangement is that of figure 15.1. Each separate winding is called a phase. In the diagram of induced e.m.f.s the arrows indicate the positive direction of the e.m.f. for each particular phase. If all three phases were kept separate a six-conductor line between source

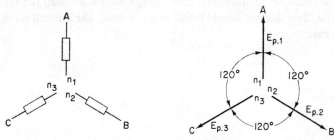

Figure 15.1   Three-phase e.m.f's.

and load would be necessary. By interconnecting the phases the number can be reduced to three; in one special case four. There are two methods.

Star connection

The three points n_1, n_2 and n_3 are joined together to form a common star or neutral point. This brings the voltage phasors of figure 15.2 to a common point n. The line conductors are connected to the

190

terminals A, B and C, so that the line voltages are E_{AB}, E_{BC} and E_{CA}. When going from A to B via the voltage phasors, the voltage E_{na} acts in the reverse direction.

$$\therefore \ E_{AB} = -E_{nA} + E_{nB} \text{ (phasor sum)}$$
$$= E_{nA} \text{ (reversed)} + E_{nB} \text{ (phasor sum)}$$

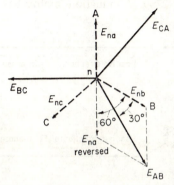

Figure 15.2   Phase and line voltages in star-connected system.

Denoting the numerical value of the line voltage by E and of the phase voltage by E_p we see that

$$E = \sqrt{3}E_p$$
$$\therefore \ E_{ab} = \sqrt{3}E_{nb} \left| -30° \right.$$
$$\text{similarly } E_{bc} = \sqrt{3}E_{nc} \left| -30° \right.$$
$$\text{and } E_{ca} = \sqrt{3}E_{na} \left| -30° \right.$$

The opposition of the positive directions of the phase e.m.f. when referred to the external circuit is also illustrated in figure 15.3. We also see that the line current is the same as the current in the phase to which it is connected. Hence, summarising

$$\text{line voltage} = \sqrt{3} \times \text{phase voltage}$$
$$\text{line current} = \text{phase current}$$

The power is given by

$$P = 3 \times \text{power per phase}$$

The power factor is the cosine of the angle between the phasors of e.m.f. and current per phase, not line voltage and current. Hence, let I_p lag (or lead) E_p by ϕ, then

$$P = 3E_p I_p \cos \phi$$
$$\text{But } E_p = E/\sqrt{3} \text{ and } I_p = I$$
$$\therefore \quad P = \sqrt{3}EI \cos \phi$$

Example 15.1.   A 2200 V induction motor has an efficiency of

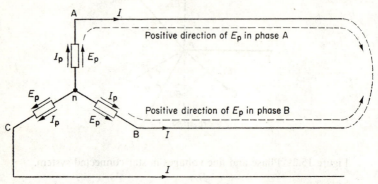

Figure 15.3   Relationship between phase and line quantities.

92 per cent and a power factor of 89 per cent when its output is 186.5 kW. Calculate the current in the line supplying the motor.

$$\text{Intake } P = \text{Output/Efficiency}$$
$$= 186.5/0.92 = 203 \text{ kW} = 2.03 \times 10^5 \text{ W}$$

For a consuming device we use V instead of E.

$$\therefore \quad P = \sqrt{3}VI \cos \phi$$
$$I = \frac{P}{\sqrt{3}V \cos \phi} = \frac{2.03 \times 10^5}{\sqrt{3} \times 2200 \times 0.89}$$
$$= 60 \text{ A}$$

Mesh, or Delta connection

The three phase windings are joined together to form a closed circuit, as in figure 15.4. Since $E_{p1} = E_{p2} = E_{p3}$ and they are mutually 120° apart their resultant, with respect to the closed circuit is zero,

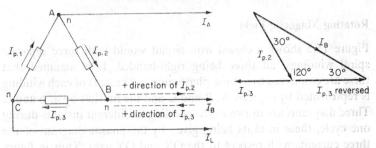

Figure 15.4   Current relationships in a mesh connected system.

and consequently no current will flow unless there is connection to an external circuit. It is quite clear that, in this case line voltage and phase voltage are equal.

$$E = E_p$$

Taking the same positive direction as before we see that for line B the directions for phase currents I_{p2} and I_{p3} are in opposition

$$\therefore \text{ Line current } I_B = I_{p2} - I_{p3}$$
$$= I_{p2} + I_{p3} \text{ reversed, phasor sum}$$

Hence with a balanced system in which all the phase currents are numerically equal and all the line currents numerically equal

$$I = \sqrt{3} I_p$$
$$\text{As before } P = 3 E_p I_p \cos \phi$$
$$= 3 E \times (I/\sqrt{3}) \cos \phi$$
$$= \sqrt{3} E I \cos \phi$$

Example 15.2.   A 3-phase consuming device with delta connected windings has an intake of 17.5 kW at a power factor of 0.85 when connected to a 230 V supply. Calculate the phase current

$$\text{Line current } I = P/\sqrt{3} E \cos \phi$$
$$= \frac{17\,500}{\sqrt{3} \times 230 \times 0.85}$$
$$= 51.8 \text{ A}$$
$$\therefore \quad I_p = 51.8/\sqrt{3} = 30 \text{ A}$$

Rotating Magnetic Field

Figure 15.5 shows a closed iron circuit would with three separate spiral windings, all three being right-handed. It is assumed that these are connected to a three-phase supply. The start of each winding is represented by capitals A, B and C and the finishes by a, b and c. Three diagrams are drawn corresponding to different instants during one cycle, these instants being given by the phasor diagram of the three currents with respect to the OX and OY axes. Thus in figure

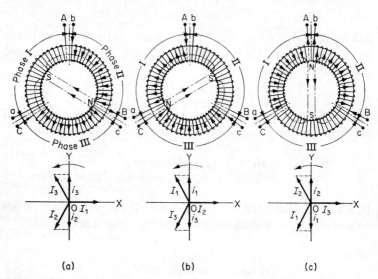

Figure 15.5 Production of a rotating magnetic field by three-phase currents.

15.5a the phasor diagram is drawn with the I_1 phasor along OX and at this instant its projection on OY is zero. Therefore the instantaneous current i_1 is zero. At the same instant, the current i_2 is negative and i_3 positive, and these two currents are numerically equal.

We will now adopt the convention that when a current is positive it enters the winding at the start (A, B or C) and leaves at the finish (a, b or c). When it is negative it enters the winding at the finish and leaves at the start. Hence, at the instant considered ($i_1 = 0$) the currents in phases two and three correspond to the arrows on the winding turns. Using the right-handed screw rule we see that phase 2 produces a m.m.f. which acts round the core in a clockwise direction, while phase 3 produces an m.m.f. which cuts in a counterclockwise direction. These m.m.f.s therefore meet at the junction of phases 2 and 3, and separate at the middle point of phase 1. Consequently the fluxes set up leave the core at this junction, cross the diagonal, as shown, and enter the core again at the middle point of phase 1. Thus the 3-phase winding has produced, externally to the core, a 2-pole magnetic field with the direction and polarity shown in figure 15.5a.

Figure 15.5b shows the state of affairs one-third of the periodic time later than figure 15.5a. The phasor diagram has rotated through 120° and the I_2 phasor now points along OX. Hence i_1 is positive i_2 is zero and i_3 is negative; and $i_1 = i_3$. The direction of the currents and fluxes are again inserted, and we see that, at this instant, the external flux is directed towards the centre point of phase 2. Thus during one-third of the periodic time the field has rotated through 120° in a clockwise direction.

Figure 15.5c shows the conditions one-third of the periodic time later than figure 15.5b. We now have for the instantaneous currents i_1 negative, i_2 positive, i_3 zero, and the field is now directed towards the centre of phase 3. After another one-third period, that is one whole periodic time, the conditions are back to those of figure 15.5a, showing that during the periodic time T the 2-pole field has made one complete revolution. Hence speed of a 2-pole field

$$N_s = 1/T = f \text{ rev/s}$$
$$= \frac{2f}{2} \text{ rev/s}$$

Windings can be arranged to produce any even number of poles. Suppose there are four poles, then the two north poles are diametrically opposite and so are the two south poles. Consequently the pattern of magnetic flux in space will be repeated after only one-half of a revolution and therefore in $T/2$ or $1/2f$ seconds.

Hence speed of a four-pole field

$$= \frac{1}{2} \times \frac{2f}{2} = \frac{2f}{4} \text{ rev/s}$$

Similarly speed of a 6-pole field $= 2f/6$ rev/s. In general the speed of a P pole field is

$N_s = 2f/P$ or f/p rev/s, where $p = P/2$ the number of pole pairs. This speed is called the synchronous speed. It is the speed of alternators and synchronous motors, and the industrially important induction motor runs at a speed only slightly less than this unless it is controlled by some speed regulating device. For a supply frequency of 50 Hz the possible synchronous speeds are therefore as follows.

No. of poles P	Pairs of poles p	rev/s	Difference
2	1	50	
			25
4	2	25	
			8.4
6	3	16.6	
			4.1
8	4	12.5	
			2.5
10	5	10	
			1.7
12	6	8.3	
			1.17
14	7	7.13	
			0.78
16	8	6.25	
			0.69
18	9	5.5	
			0.56
20	10	5	
			0.45
22	11	4.55	
			0.38
24	12	4.16	

Since there are great practical advantages in the use of high-speed machines it follows that the gap of 25 between the 2-pole and 4-pole speeds, and the gap of 8.4 between the 4-pole and 6-pole speeds are of considerable importance. They show the very limited number of available speeds at the most important part of the range.

16 THE TRANSFORMER

The transformer is a stationary appliance used to change the voltage of an a.c. supply without changing the frequency. For long distance transmission very high voltages are required; for distribution, much lower voltage; and for utilisation, say in the home, voltages of the order of 240 V. The transformer is essentially a 'close-coupled' mutual inductor, this meaning that almost of the whole of the flux set up by the primary m.m.f. links with the secondary winding, and almost of the whole of the flux set up by the secondary m.m.f. links with the primary winding. In other words, the leakage flux, the flux which links with one winding only, is very small. To ensure such close coupling the windings are arranged with their coils in close physical association and they are wound on a magnetic core.

The transformer therefore consists essentially of (a) the core; (b) the windings, and their insulation from one another and from the core.

The Core

There are three alternative methods of core construction, namely, core type, shell type and distributed core type. In all cases the core is laminated for reasons given below. Figure 16.1 shows the three methods of construction. Figure 16.1a shows the core type which consists of two vertical limbs and two horizontal yokes. The windings are placed on the limbs only and there is a single magnetic circuit, shown dotted. Figure 16.1b shows the shell type; the winding is placed on the central limb which is of double the cross-section of the horizontal and vertical portions of the yokes. This is because the transformer has a double magnetic circuit, shown dotted. Figure 16.1c shows the distributed core construction, often used, because of its compactness, for small pole-mounted distribution transformers.

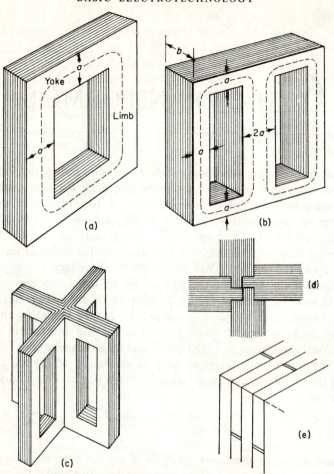

Figure 16.1   Core construction for single-phase transformer.

Figure 16.1d gives the detail of the core joint for the core of figure 16.1c. Figure 16.1e shows the interleaved or 'imbricated' joint used at the corners of the core so as to ensure a low magnetic reluctance there. With large transformers two or more widths of strip are used when building up the core on which the windings are placed, but not the yokes. This is to obtain a section which fits more closely into a circle, this being the figure which has the smallest circumference for a given area, and which therefore requires the smallest possible length of conductor in the windings. In the shell-type core both limb and yokes are almost invariably rectangular because the dimension marked *b* is anything up to 4 times the dimension marked *a*, and therefore the approximate circular section is not possible.

Transformers are 'step-up' or 'step-down' depending on whether the secondary induced e.m.f. is greater than, or less than, the primary applied p.d.

The Windings and Insulation

Core Type. The coils are in the form of concentric cyinders, the high-tension (HT) being outside the low-tension (LT). Each limb

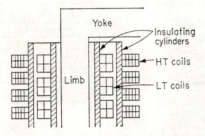

Figure 16.2   Core-type transformer construction.

has one-half of the HT and one-half of the LT turns, although for the sake of simplicity, diagrams show all the HT turns on one limb and all the LT turns on the other. Each cylinder is divided into sections which are kept apart by small radial tabs. This provides spaces between the sections for the free flow of the oil in which the transformer is immersed. The conductor is copper, round wire for small currents and rectangular strap for heavy currents. Except for small transformers the conductor covering is paper strip wound on half lap. The paper used is mechanically strong and is a good heat conductor. The insulation between HT and LT cylinders and between the windings as a whole and the core is in the form of cylinders of pressboard, a fibrous material which becomes impregnated with transformer oil. The construction is shown in figure 16.2. The insulation of the conductors themselves is the 'minor' insulation. The insulating cylinders are the 'major' insulation.

Shell Type. Instead of having coils in the form of long cylinders with several turns per layer, the shell type has thin coils, often with only one coil per layer. The conductor is rectangular strap wound on the flat with a strip of fibrous material wound on at the same time to provide the insulation between turns. As the core section is rectangular (figure 16.1b) each coil is rectangular in form, and because of its

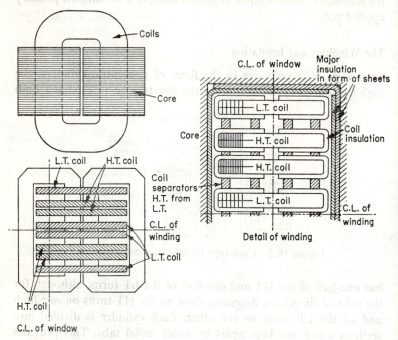

Figure 16.3   Construction of single-phase shell-type transformer.

thinness is often called a pancake coil. Unlike the core type, the HT coils are interleaved with the LT the coil groups being so arranged that there is a half group of LT coils at each end, so as to facilitate the insulation from the core. The major insulation is in the form of sheets and spacing strips of fibrous insulations, a typical construction being shown in figure 16.3.

Losses of power in the core

Since the core is subjected to alternating cycles of magnetism there is an inevitable loss of power, due to hysteresis. It is to keep this loss to an acceptable minimum that high grade materials such as silicon sheet steel are used for the core. There is an additional loss, the eddy-current loss. Figure 16.4a shows a magnetic core with a magnetising winding, the circular current flow producing an axial flux. If the core carries an alternating flux from an a.c. magnetising coil (not shown), then a circular turn in the form of a closed ring will have a current induced in it, and there will be an I^2R loss, figure 16.4b. But the core itself is a conductor and we can imagine

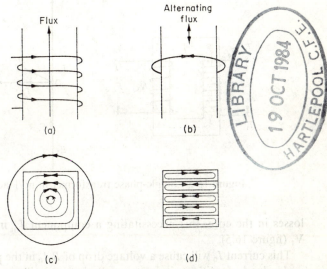

Figure 16.4 Production of eddy-currents.

that it is subdivided into elementary induced current paths, as in figure 16.4c. With a solid core the resulting I^2R loss will be so great that, besides being unacceptably high, it will raise the temperature to such an extent that the device will be unworkable. To prevent this the core is laminated, thereby dividing the eddy-current paths into a number of paths of higher resistance, and in which the e.m.f. per path is very small (figure 16.4d). The usual thickness of the sheet steel is about 0.4 mm. Lamination does not reduce the hysteresis loss.

The phasor diagram

First of all consider the transformer on no-load. The secondary will have induced in it the secondary no-load e.m.f. $E_{2.0}$, but, carrying no current it will have no influence on the primary, which will therefore act like an inductor of very high self inductance. The applied primary p.d. V_1 will produce the primary no-load current I_0. This will have two things to do: first to produce the magnetic flux Φ, necessitating a magnetising component I_μ in quadrature lagging V_1; second, provide the power for the combined hysteresis and eddy-current

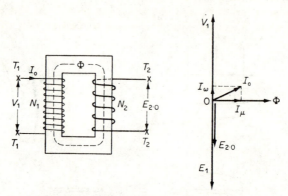

Figure 16.5   Single-phase transformer in no load.

losses in the core, this necessitating a component I_ω in phase with V_1 (figure 16.5).

This current I_0 will cause a voltage drop of $R_1 I_0$ in the primary but, under load conditions it is so small that we will neglect it. The alternating flux will therefore induce in the primary an e.m.f. E_1 equal to, and in phase opposition to V_1. It will also induce the secondary e.m.f. $E_{2.0}$ in phase with E_1. In the figure the phasor diagram is drawn for a step-down transformer with $E_{2.0}/V_1 = 1/2$. Since the same alternating flux links with all the turns of both windings, it follows that the volts per turn, E_t, will be the same for both

$$\therefore \quad V_1 \simeq E_1 = N_1 E_t$$
$$E_{2.0} = N_2 E_t$$
$$\therefore \quad E_{2.0}/E_1 = N_2/N_1$$

This shows that the voltage ratio at no-load is equal to the turns ratio.

Now let the secondary winding deliver a load current of I_2 at a

THE TRANSFORMER 203

power factor of cos ϕ_2, as in figure 16.6. Because of internal volt-drops from the winding resistances R_1 and R_2 and also from reactances resulting from fluxes which link with one winding only the secondary terminal p.d. V_2 on load will be less than $E_{2.0}$ and also not quite in phase with it although, for simplicity, this is not shown in the phasor diagram. The current I_2 will result in a secondary m.m.f. of N_2I_2 and this, in turn, will set up a secondary flux Φ_2 taking the main magnetic path. This upsets the magnetic balance of the no-load conditions and induces a current I_1' in the primary winding which is closed through the source. This current will be in

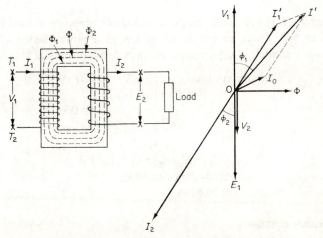

Figure 16.6 Single-phase transformer in load.

phase opposition to I_2. This current, in turn, sets up a primary m.m.f. of N_1I_1' which produces a third flux Φ_1 also taking the same magnetic path. For steady-state conditions these fluxes Φ_2 and Φ_1 neutralise one another so that, even on load, the total flux remains at the original no-load value Φ. The total primary current I_1 is the phasor sum of the induced primary current I_1' and the no-load current I_0. Thus, in general $\phi_1 \neq \phi_2$. Except at fractional loads the difference is small because I_0 is small and we can therefore assume that the power factors of both sides are the same.

Two important results from the fact that the fluxes Φ_1 and Φ_2 do not exist under steady-state conditions:

1. The total flux remains at Φ whether the transformer is loaded

or not. Hence the transformer supplied at constant primary applied p.d. is a constant flux device and therefore the core loss is constant.

2. The secondary m.m.f. and counterbalancing primary m.m.f. are equal and opposite at every instant

$$N_1 I_1' = N_2 I_2$$
$$\therefore \quad N_1 I_1 \simeq N_2 I_2$$
$$I_2 / I_1 \simeq N_1 / N_2$$

the current ratio therefore being the inverse of the turns ratio. Hence if a transformer steps down the voltage it steps up the current, and *vice versa*.

Example 16.1.   A single-phase transformer stepping down from 3000 to 600 V has an output of 50 kW at a power factor of 0.8 lagging. Calculate the primary and secondary currents at this load.

$$\text{Secondary load } P_2 = 50 \text{ kW} = 50\,000 \text{ W}$$
$$\text{But } P_2 = V_2 I_2 \cos \phi$$
$$\therefore \quad I_2 = \frac{P_2}{V_2 \cos \phi} = \frac{50\,000}{600 \times 0.8}$$
$$= 104.2 \text{ A}$$
$$I_1 \simeq I_2 \times (N_2 / N_1) \simeq I_2 \times (V_2 / V_1)$$
$$\simeq 104.2 \times (600 / 3000)$$
$$\simeq 20.84 \text{ A}$$

The e.m.f. equation

The flux alternates between Φ_m and $-\Phi_m$, the change therefore being $2\,\Phi_m$. This change takes place in one-half of the periodic time, that is in $T/2$ second. With a frequency of f hertz

$$T/2 = 1/(2f)$$

$\therefore$ Average induced e.m.f. per turn

$$E_t = \frac{\text{change of flux}}{\text{time of change}}$$
$$= \frac{2\,\Phi_m}{1/(2f)} = 4\,\Phi_m f$$

Hence for a winding of N turns

$$E_{av} = 4\,\Phi_m N f$$

For a sinusoidal e.m.f. we have seen that the r.m.s. value, E, is 1.11 times the average value

$$\therefore \ E = 4.44 \ \Phi_m N f$$

Example 16.2. A 100 W transformer for a radio receiver power pack has a net iron core section of 16 cm². The core is worked at a maximum flux density of 1.0 T. How many primary turns are required for a supply at 240 V, 50 Hz.

$$\Phi_m = B_m a = 1 \times 16 \times 10^{-4} = 1.6 \times 10^{-3} \ \text{Wb}$$

E.m.f. per turn

$$E_t = 4.44 \ \Phi_m f = 4.44 \times 1.6 \times 10^{-3} \times 50$$
$$= 0.352 \ \text{V}$$

$$\therefore \ N_1 = V_1/E_t = 240/0.352 = 682 \ \text{to the nearest whole turn.}$$

Efficiency

The efficiency of any machine can be written

$$\text{Efficiency, } \eta = \frac{\text{Output}}{\text{Intake}}$$

$$= \frac{\text{Output}}{\text{Output} + \text{Losses}}$$

The losses of power in a transformer are the I^2R losses in the windings; the hysteresis and eddy-current losses in the core, known collectively as the core loss or iron loss; and the additional load losses. These are losses due to currents induced by leakage fluxes in all the metalwork they penetrate; also to some extent in the conductors themselves if these are of large cross-section. Denoting the iron losses by P_i and the additional load loss by P_a, the total losses are

$$R_1 I_1^2 + R_2 I^2 + P_i + P_a$$

The output is

$$V_2 I_2 \cos \phi$$

$$\therefore \ \eta = \frac{V_2 I_2 \cos \phi}{V_2 I_2 \cos \phi + R_1 I_1^2 + R_2 I_2^2 + P_i + P_a}$$

Example 16.3. A 15 kVA transformer has a no-load voltage ratio of 2300/240 V, but when delivering the full secondary current at a power factor of 0.8 lagging the secondary terminal p.d. is 230 V.

Calculate its efficiency given an iron loss of 100 W, negligible additional load loss, $R_1 = 2.7 \ \Omega$ and $R_2 = 0.025 \ \Omega$

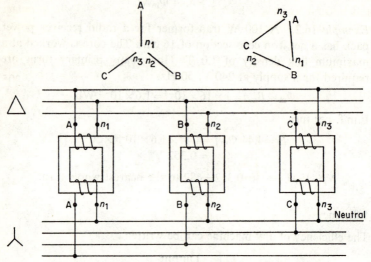

Figure 16.7   Three single-phase transformers connected.

$$I_2 = \text{Volt-amperes}/V_2$$
$$= 15\,000/240 = 62.5 \ \text{A}$$
$$I_1 \simeq 62.5 \times (240/2300)$$
$$\simeq 6.5 \ \text{A}$$
$$\therefore \quad R_1 I_1^2 \simeq 2.7 \times 6.5^2 = 112 \ \text{W}$$
$$R_2 I_2 = 0.025 \times 62.5^2 = 98 \ \text{W}$$
$$\therefore \ \text{Total losses} = 112 + 98 + 100 = 310 \ \text{W}$$
$$\text{Output} \quad = 240 \times 62.5 \times 0.8 = 12\,000 \ \text{W}$$
$$\therefore \ \text{Efficiency} \quad = \frac{12\,000}{12\,000 + 310} = 97.4 \ \text{per cent}$$

Three-phase transformation. Four-wire distribution

Three-phase transformation can be effected by a 'bank' of three separate single-phase transformers, or by a single three-phase unit. The windings of both primary and secondary sides can be connected in star or in delta, and it is not necessary that the system should be the same on both sides. Thus the possible arrangements are Y—Y, Δ—Δ, Y—Δ and Δ—Y, as in figure 16.7.

A three-phase single unit is almost invariably of the core type. It will be seen from figure 16.8 that the magnetic circuit consists of three limbs, all of the same cross-section, joined together at the top and bottom by yokes. Each limb carries the whole of the primary and secondary turns of one phase. Figure 16.7 shows Δ—Y connections, a common arrangement with step down transformers since it gives on the secondary side a neutral point which can be earth connected, thereby fixing the potentials with respect to earth of

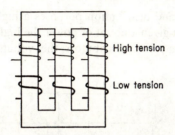

High tension

Low tension

Figure 16.8   Three-phase core-type transformer.

everything connected to this side. It also enables a fourth wire for four-wire distribution, as explained below.

Example 16.4.   A three-phase transformer steps down from 3300 to 600 V (line voltages). It is connected Δ—Y. If its output is 200 kVA, power factor 0.7, and its efficiency at this loading is 0.97, calculate the line currents and line voltages, and phase currents and phase voltages on both HT and LT sides.

Output in kW = kVA × p.f.

$$\therefore P_2 = 200 \times 0.7 = 140 \text{ kW}$$

$$I_2 = \frac{P_2}{\sqrt{3}V_2 \cos \phi} = \frac{140\ 000}{\sqrt{3} \times 600 \times 0.7} = 192 \text{ A}$$

This is the line current, and as the secondary side is star connected it is also the phase current.

Phase voltage in secondary side $V_{p.2} = 600/\sqrt{3} = 347$ V. Intake on primary side

$$P_1 = P_2/\eta = 140/0.97 = 144.5 \text{ kW}$$

$$\therefore I_1 = \frac{P_1}{\sqrt{3}V_1 \cos \phi} = \frac{144\ 500}{\sqrt{3} \times 3300 \times 0.7} = 36.15 \text{ A}$$

This is the line current at the primary side. Hence with delta connection

$$I_{p.1} = 36.15/\sqrt{3} = 20.8 \text{ A}$$

and primary phase voltage is equal to primary line voltage

$$V_{p.1} = 3300 \text{ V}$$

Four-wire distribution

For transmission and distribution purposes a high voltage is desirable because, for a given amount of power the higher the voltage the smaller the current. For distributors the voltage is fixed by the

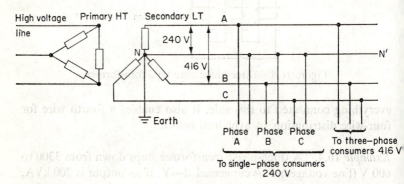

Figure 16.9   Three-phase 4-wire distributor with both single-phase and three-phase loads.

requirements of the consuming devices, the lowest utilisation voltage being that in the home, 240 V. By connecting single-phase consuming devices between one line and the neutral of a star-connected system the line can be operated at $\sqrt{3}$ times that of the consumer voltage, namely $\sqrt{3} \times 240 = 416$ V. At the same time, devices which can be conveniently worked at the higher voltage, for example, three-phase motors, can be connected to the line conductors. The arrangement is thus that of figure 16.9. If the single-phase loads could be equally divided between the three phases, both as to kVA and power factor the neutral current would be zero. This ideal is, of course, not possible but unless the out of balance is large the current in the neutral wire will be small and therefore the volt-drop in it will be small. This means that the voltage at N′ the far end of the neutral wire will be very little different from that at N. Hence if N is connected to earth then N′ will also assume a potential very close to that of earth, and the three live wires will be maintained at 240 V relative to earth. This holding of all the potentials with respect to earth by the earthing of the neutral point of the transformer secondary is a very important advantage of the system.

Tap changing

With fixed ratio transformers there is no control over the secondary terminal voltage of a transformer, and therefore, because of the volt drops in the line supplying it, there will be a voltage diminution at

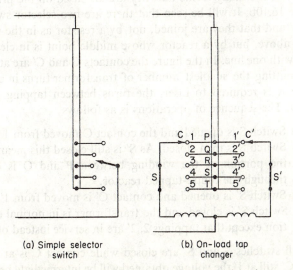

(a) Simple selector switch

(b) On-load tap changer

Figure 16.10   Principle of on-load tap changing.

the consumer terminals in times of heavy load, and voltage fluctuations when the load changes suddenly. By arranging tappings on one side of the transformer so that the number of turns in circuit can be increased or decreased as required, the above difficulty can be overcome. The simplest arrangement is a selector switch as shown in figure 16.10a. Its disadvantages are first, if the selector switch wipes only one contact at a time the circuit will be opened when it is changed from one contact to the next, and second, if it is made to bridge two contacts then, when it is changed from one contact to another this bridging will take place and so short-circuit a portion of the winding. For small transformers this can be overcome by adopting the device used on battery switches. The moving contact consists of two contacts joined electrically by a resistor. At the transition position between two permanent positions the portions of the winding bridged by the two movable contacts will then be closed through the resistor.

For all but small transformers the above simple method is not suitable and tap-changers have been developed which neither interrupt the current nor risk damage to the transformer or to the tap-changer contacts. Furthermore they can be operated automatically. There are many forms but they are all based on the principle of figure 16.10b. It will be seen that there are two selector switches S and S' and that they are joined, not by a resistor as in the simple method above, but by a reactor whose middle point is in electrical contact with one line. In the figure the contacts C and C' are at 1 and 1' thus putting the smallest number of transformer turns in circuit. Suppose it is required to insert the turns between tapping points P and Q. The sequence of operations is as follows

   (i) Switch S is opened and the contact C moved from 1 to 2.
  (ii) Switch S is then closed. As S' is still closed this means that the portion of the winding between P and Q is closed through the centre-tapped reactor.
 (iii) Switch S' is opened and contact C' is moved from 1' to 2'.
  (iv) Switch S' is closed and the transformer is in normal operation except that tappings 2, 2' are in service instead of 1, 1'.

If both switches S and S' are closed while contact C is at 2 and contact C' still at 1' the voltage obtained will be intermediate between the voltages at 1, 1' and 2, 2'. This means that the reactor must be 'continuously rated' but it doubles the number of voltage steps for a given number of tappings.

The discussion of other methods and of the best location for the tap changer are beyond the scope of this book but it can be said that, in general, the HT side is preferable because the current-carrying parts are smaller and the gear therefore less bulky and expensive. Also if located on the LT side this winding must be on the outside and the HT winding thereby adjacent to the limb.

Regulation

By regulation we mean the change in secondary terminal voltage from no-load to full-load. Thus if $E_{2.0}$ is the secondary e.m.f. on no-load and V_2 the secondary terminal p.d. on full-load at some stated power factor, then

$$\text{Regulation} = E_{2.0} - V_2$$
$$\text{and percentage regulation} = 100(E_{2.0} - V_2)/E_{2.0}$$

The internal volt-drop is due both to ohmic resistance and inductive reactance. Consider first of all the resistance

$$\text{Primary ohmic drop} = R_1 I_1$$
$$\text{Now } I_1 \simeq I_2 \times (N_2/N_1)$$
$$\simeq I_2 \times (E_{2.0}/V_1)$$

If this is transformed the corresponding secondary current will be

$$I_2 \simeq I_1 \times (V_1/E_{2.0})$$
$$\therefore \text{ The loss of power in the primary}$$
$$R_1 I_1^2 \simeq [(E_{2.0}/V_1)^2 \; R_1] I_2^2$$

The expression in the square brackets is the primary resistance referred to the secondary. Denoting it by $R_{1.2}$ we have

$$R_{1.2} = (E_{2.0}/V_1)^2 R_1$$

The primary volt-drop due to resistance, when referred to the secondary is therefore

$$R_{1.2} I_2$$

and the total volt-drop is

$$(R_{1.2} + R_2) I_2$$

Example 16.5   A 50kVA transformer which steps down from 6600 to 220 volts has $R_1 = 7.8$ and $R_2 = 0.0085 \; \Omega$ respectively. Calculate the resistance referred to the secondary, and hence calculate (a) the total volt-drop due to resistance, (b) the full-load copper loss

$$\text{Full-load secondary current } I_2 = \text{volt amps}/E_{2.0}$$
$$= 50\,000/220 = 227 \text{ A}$$
$$R_{1.2} = (220/6600)^2 \times 7.8 = 0.00866$$
$$\therefore \; R_{1.2} + R_2 = 0.00866 + 0.0085 = 0.0176 \; \Omega$$
$$\text{Full-load volt-drop due to resistance} = 0.01716 \times 227 = 3.9 \text{ V}$$
$$\text{Full-load copper loss} = 0.01716 \times 227^2 = 885 \text{ W}$$

The inductive reactance of the windings is not due to the total flux, which links with both primary and secondary, but to the leakage fluxes which link with one winding only. Suppose that this leakage reactance, referred to the secondary winding is 0.08Ω, then the full-load volt-drop due to it is

$$0.08 \times 227 \simeq 18.2 \text{ V}$$

Now the resistance drop is in phase with I_2, but the reactive drop is in quadrature leading I_2. Suppose that I_2 is in phase with V_2 then

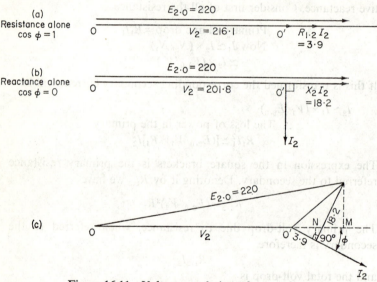

Figure 16.11   Voltage regulation of a transformer.

the phasor diagram for the secondary side is that of figure 16.11a. We see that, numerically, the drop is due almost entirely to the resistance, giving

$$V_2 \simeq 220 - 3.9 = 216.1 \text{ V}$$

Now suppose that I_2 is in quadrature lagging V_2, figure 16.11b, then the reactive drop $X_2 I_2$ is now in direct opposition to $E_{2.0}$, and we now have

$$V_2 \simeq 220 - 18.2 = 201.8 \text{ V}$$

At a power-factor in between 1.0 and zero both components produce appreciable volt-drops, as shown in figure 16.11c. Suppose that $\cos \phi = 0.8$, so that $\sin \phi = 0.6$. The angle at O, if the phasor diagram is drawn to scale will be very small, so that

$$E_{2.0} \simeq \text{OM} = V_2 + 3.9 \cos \phi + 18.2 \sin \phi$$
$$220 = V_2 + 3.9 \times 0.8 + 18.2 \times 0.6$$
$$= V_2 + 14$$
$$\therefore \quad V_2 = 220 - 14 = 206 \text{ V}$$

and percentage regulation $= \dfrac{14}{220} \times 100 = 6.37$ per cent

at 0.8 p.f. loading.

17 ELECTRIC MACHINES

The function of all rotating electric machines is the conversion of one form of energy to another: mechanical into electrical energy in the case of generators; electrical into mechanical energy in the case of motors. The fundamental facts which make this conversion possible are first that an e.m.f. is induced in a moving conductor when it cuts the lines of force of a magnetic field, or when the number of lines of force cutting a circuit changes; and second, that a conductor carrying current is acted on by a mechanical force if it distorts the magnetic field in which it is placed. We see from the first consideration that an electric machine must consist essentially of two parts, namely that which produces the magnetic field, and that which carries the conductors in which the e.m.f.s are induced. It is obvious that there must be relative motion between these two parts.

In direct-current machines the field system is stationary and the

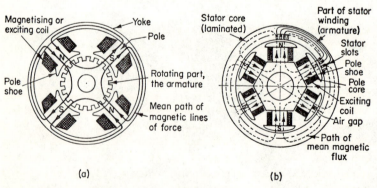

Figure 17.1 Arrangements for magnetic circuits of direct-current and salient-pole alternating current machines.

214

armature, the part in which the e.m.f.s are induced, rotates inside it. In alternating-current machines the reverse arrangement is almost invariable. This is because the e.m.f. induced in the armature winding may be as high as 33 000 V and clearly a large current at such a high voltage could not be collected from a rotating device. The armature winding is therefore housed in a stationary member, the stator. The field-magnet system, the rotor, thus rotates inside the stator which

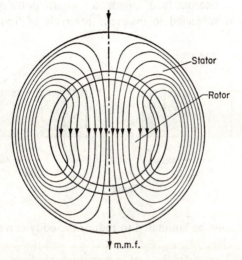

Figure 17.2 Magnetic flux distribution of two-pole machine with cylindrical rotor slots in which the windings are housed (not shown).

means that its current must be supplied via two slip rings. As the power taken by the field winding is small and the voltage low, about 100 V, no difficulty is experienced.

The magnetic circuits of d.c. machines and salient pole a.c. machines are compared in figure 17.1. The turbine-driven alternator has a cylindrical rotor with a field winding housed in slots. The air-gap is thus of constant radial length and the flux paths for a two-pole machine are as indicated in figure 17.2. It follows from the above that the magnetic field of the d.c. machine is stationary in space. Because of its direct-current excitation the magnetic field is fixed with respect to the rotor structure and therefore rotates with it. Hence the magnetic field of the a.c. machine rotates in space at synchronous speed N_s, since it is this speed which fixes the frequency $f = N_s p$ where $p =$ number of pole pairs.

The a.c. Synchronous machine

Figure 17.2 shows the flux distribution of a two-pole machine at one particular instant of time. Although the stator structure is stationary, the magnetic flux rotates in space so that any given region of the stator face is alternately of north and south polarity. Hence the stator core is subjected to magnetic reversals of frequency f, and

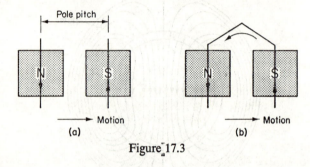

Figure 17.3

therefore it must be laminated to reduce the eddy-current loss to an acceptable value.

Figure 17.3 provides the key to armature windings, both a.c. and d.c. It shows two conductors moving from left to right in front of two poles, or alternatively, the two poles moving from right to left behind two stationary conductors. Figure 17.3a shows that the induced e.m.f.s are in opposite directions in space. Figure 17.3b shows that when two conductors are joined to form a coil this condition must apply in order that the e.m.f.s act together in the coil. Hence the coil width should be equal to, or not very different from, the pole pitch. Figure 17.3b shows the simplest possible single-phase winding, an open coil, stationary in space, its ends taken to fixed terminals.

Such a winding is impracticable since first, with only one slot per pole there is very poor utilisation of the active materials, and second, to hold all the conductors required the slots would be much too big. For these reasons the winding consists of several coils *distributed* over the stator face. Figure 17.4 shows two shapes of coil for a

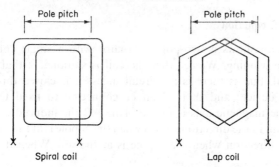

Figure 17.4 Coils distributed over three slots per pole.

winding distributed over three slots per pole. The individual coils can, of course, consist of several turns. The four-pole winding is two two-pole windings in series, and similarly for other numbers of poles, showing that the operation of the machine can be studied by considering the two-pole case.

Figure 17.5 shows the simplest possible three-phase winding. For

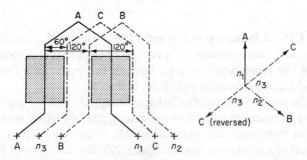

Figure 17.5 Two-pole three-phase winding with one slot per phase per pole.

purely mechanical convenience the winding of the C phase is placed 60° from the A phase and therefore, for connection to the external circuit, its terminals C and n_3 have to be taken in reverse order. As with the single-phase winding, a practical three-phase winding is distributed over several slots per phase per pole.

B.E.—8

The e.m.f. equation

Figure 17.6 shows a two-pole alternator with a single coil for its armature winding. We see that the coil is stationary; that it is an open coil, that is there is no circuit unless it is closed through an external device; and that it can be connected to fixed terminals. When the magnetic axis is in the plane AB the induced e.m.f. is a maximum, in the direction shown when the N pole is at the top; in the reversed direction when the S pole is at the top. When it is in the

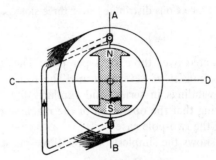

Figure 17.6   Simple single-phase 2-pole alternator.

plane CD the induced e.m.f. is zero. Hence the e.m.f. goes through one cycle in one revolution of a two-pole field and it is obvious that during one revolution it will go through two cycles in a four-pole field, three cycles in a six-pole field, and so on.

In the position CD the flux linkage through the coil is a maximum, namely Φ, where Φ is the flux per pole. If the coil has N turns then the flux linkage in this position is ΦN. When the coil has made one-half of a revolution, that is, after $1/2f$ s, the flux through the coil is reversed. Hence

$$\text{Change in flux linkage} = 2\Phi N$$
$$\text{Time of change} = 1/2f$$
$$\therefore \quad E_{\text{av}} = 2\Phi N \div 1/2f$$
$$= 4\Phi Nf$$
$$\therefore \quad E = 1.11 E_{\text{av}}$$
$$E = 4.44\Phi Nf$$

It is sometimes convenient to think in terms of conductors instead of in coils, and since each coil has two sides the induced e.m.f. in terms of the number of conductors Z

$$E = 2.22\Phi Zf$$

A correction has to be applied (a) because the flux density distribution round the air-gap may not be sinusoidal, in which case the e.m.f. per conductor will not be sinusoidal. (b) with a distributed winding in which there will be a small phase difference between the e.m.f. in

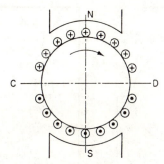

Figure 17.7 Distribution in space of the induced e.m.f.'s. in a 2-pole d.c. machine.

each coil so that their phasor sum is less than their arithmetic sum. We therefore introduce a constant k, giving

$$E = 2k\Phi Zf \text{ for a single-phase winding, and}$$
$$E_{\text{ph}} = 2k\Phi Z_{\text{ph}}f \text{ for a three-phase winding}$$

Example 17.1   A star-connected, 50-cycle, three-phase alternator has 108 stator slots each containing 5 conductors. If the flux per pole is 0.00636 Wb, calculate the terminal e.m.f. on no-load, given that the constant k for the winding is 1.06.

$$Z = 108 \times 5 = 540$$
$$\therefore \quad Z_{\text{ph}} = 540/3 = 180 \text{ conductors per phase}$$
$$\therefore \quad E_{\text{ph}} = 2 \times 1.06 \times 0.00636 \times 180 \times 50$$
$$= 121 \text{ V}$$
$$\therefore \quad E = \sqrt{3} \times 121 = 210 \text{ V}$$

The d.c. machine

In this machine it is the armature which rotates, but the fundamental difference is that the armature winding is closed on itself, irrespective of any connection to outside circuits. Figure 17.7 shows the distribution in space of the e.m.f.s induced in the armature of a d.c. machine. Although the armature rotates the e.m.f. distribution remains fixed in space and, as we shall see, this is the basis of what is called commutation. It is clear that the sum of the e.m.f.s of all the coils above the neutral plane CD will act in one direction round the winding,

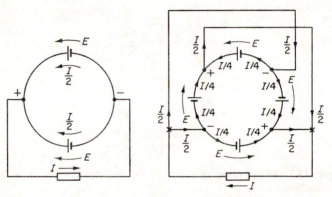

Figure 17.8   Analogues of two-pole and four-pole d.c. machines.

while the sum of the e.m.f.s in all the other conductors will act in the opposite direction round the winding. Thus there will be a balance of e.m.f.s and, in spite of the fact that the winding is closed on itself, no current will flow unless the armature is connected to an external load. In this respect the armature acts like a battery of cells in parallel. Figure 17.8 shows batteries of two and of four cells, analogous to two-pole and four-pole machines respectively. In each case there is a balance of e.m.f.s, the total e.m.f. acting round the local circuit being zero. If this was not the case, batteries of cells in parallel would be unworkable. The important thing to notice is that the positive terminals are meeting points of two e.m.f.s, while the negative terminals are separating points of two e.m.f.s.

When current is delivered to an external circuit this current is shared by a number of parallel paths, two in the case of the two-cell battery and four with the four-cell battery.

Figure 17.9 shows a four-pole armature winding housed in 14 slots and we are assuming that there are two coil sides per slot. It will be

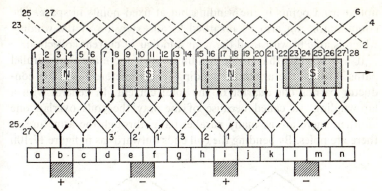

Figure 17.9   Distribution of e.m.f's. in a lap winding.

seen that the basic unit is a coil of width nearly equal to the pole pitch as with the windings of figures 17.3 and 17.5. The coils are shaped so that one coil side lies at the top of a slot, and the other side (shown dotted) at the bottom. To simplify the diagram they are all separated. It is assumed that motion of the winding is such that all the conductors under the N poles have e.m.f.s which act downward, while all those under the S poles have upward e.m.f.s. We see that there are two meeting points and two separating points so that the scheme is analogous to the four-cell arrangement of figure 17.8.

At the moment shown, point 1 is a meeting point and point 1' a separating point; when the armature has moved a distance equal to that between consecutive conductors point 2 will occupy the point 1 and point 2' will occupy the point 1', and similarly with points 3 and 3', and so on. Hence as the armature rotates the meeting and separating points of two e.m.f.s always take place at the same points in space. This is the explanation of the rectification of the alternating e.m.f.s induced in the individual conductors; it is because the magnetic field of the machine is stationary in space.

The junctions of adjacent coils are taken to the segments, a, b, c etc, of a commutator, which is a cylindrical structure consisting of copper segments insulated from one another and from the frame of the armature. Therefore the commutator is not a rectifier; it is, as its name implies, a device for bringing into electrical contact with the collecting brushes each junction of two coils at the moment that junction becomes either a meeting point or separating point of two e.m.f.s. As we have seen, these meeting and separating points occur, not at fixed points in the winding but at fixed points in space. The commutator can also be regarded as an extension of the winding which is able to withstand the wear and tear of current collection.

It will be clear that in the winding shown there are four parallel paths through the armature winding, and therefore that each conductor carries one-quarter of the total current. In general with a lap-winding, so called because of the way the coils overlap one another, there are as many parallel paths through the armature as there are poles. By bending the coil ends outwards as in figure 17.10b

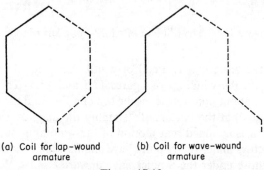

(a) Coil for lap-wound armature (b) Coil for wave-wound armature

Figure 17.10

a wave winding is obtained and in this there are only two parallel paths through the winding irrespective of the number of poles. This winding is commonly used for small four-pole machines operating at 200 V or more.

The brushes which collect the current from the commutator are invariably of carbon, the grade depending largely on the voltage of the machine.

Since there is relative motion between armature and magnetic field the armature core has to be laminated in order to reduce eddy current loss to a minimum.

e.m.f. equation

Figure 17.11 shows a conductor about to enter the field of a pole at point A, and about to leave the field at point B. In moving from A to B it will cut all the Φ lines of the flux from the pole. Hence average e.m.f. per conductor

$$E_1 = (\text{Flux cut})/\text{Time of cutting}$$

With a speed of n rev/s the time of one revolution is $1/n$ s. The distance AB is equal to $1/P$ of the circumference, where P is the number of poles. Hence time of cutting the Φ lines per pole is $1/nP$

$$\therefore \quad E_1 = \Phi/(1/nP) = \Phi nP$$

From figure 17.10 we see that the conductors are uniformly distributed over the magnetic field so that we can multiply E_1 by the number of conductors to give the total e.m.f. induced in that path

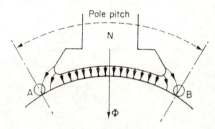

Figure 17.11

and therefore the induced e.m.f. of the machine. Let Z be the total number of conductors and A the number of parallel paths then

$$\text{Conductors per path} = Z/A$$
$$\therefore \quad E = \Phi Z n \times (P/A)$$

For a lap winding we have seen that $A = P$, giving $E = \Phi Z n$

Example 17.2 A four-pole generator has 75 slots each with 6 conductors. It is lap wound. If its speed is 600 rev/min and the flux per pole is 0.064 Wb, calculate the induced e.m.f.

$$\Phi = 0.064 \text{ Wb}, Z = 75 \times 6 = 450; n = 600/60 = 10 \text{ rev/s}$$
$$\therefore \quad E = 0.064 \times 450 \times 10 = 288 \text{ V}$$

Generator and Motor action

Work is done when force is overcome. With linear motion a force overcomes an opposing force. With rotation a torque overcomes an opposing torque. In the electric circuit an e.m.f. overcomes an opposing e.m.f. The inductor is an illustration of this third case. The opposing e.m.f. is the back e.m.f. due to the establishment of the magnetic field and the work done is converted into the stored energy of this field.

Figure 17.12a shows that the armature currents produce a magnetic field independently of the main field. In the case of a two-pole generator with N pole at the top and S pole at the bottom the current directions for clockwise rotation are as shown. The armature field acts from right to left and we can regard the armature as an electromagnet with N pole to the left and S pole to the right. Figure 17.12b

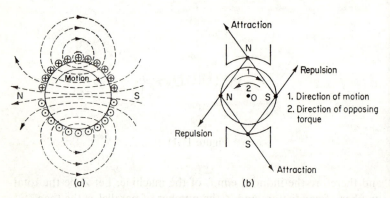

Figure 17.12   Production of the opposing torque of a generator. Arrow 1, driving torque. Arrow 2, opposing torque.

shows the armature and field magnetic systems acting together, and since like poles repel and unlike poles attract we see that a torque will be exerted on the armature, and that this torque is in opposition to motion. It is this opposing torque which has to be overcome by the prime mover driving the generator.

Now suppose that the machine is not coupled to a prime mover but that the field and armature windings receive current from an outside source. Furthermore let the polarity of the field and the directions of the armature currents be the same as in figure 17.12 then, magnetically the state of the machine will be exactly the same as in that figure. Hence, in the absence of an impressed torque the armature will rotate in the opposite direction under the influence of the torque 2. The machine will now be a motor. We see that for the same direction of the main field and the same direction of the armature currents a machine, when acting as a motor, rotates in the opposite direction to action as a generator (figure 17.13).

The conductors are now cutting the lines of force of the main field in the reverse direction and therefore, in each conductor, there will be induced an e.m.f. in opposite direction to the current, figure 17.13a. Hence the total induced e.m.f. E_b is in opposition to the applied p.d. V, and it is by opposing this e.m.f., the back e.m.f.,

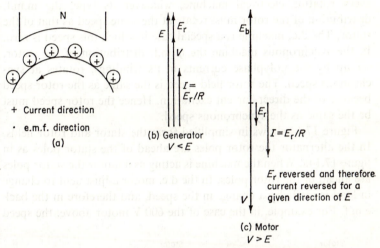

+ Current direction

· e.m.f. direction

(a)

(b) Generator

$V < E$

(c) Motor

$V > E$

E_r reversed and therefore current reversed for a given direction of E

Figure 17.13   Generating and motoring regimes contrasted.

that work is done. It follows that, in both generator and motor cases there will be a difference between the induced e.m.f. and this difference, E_r, will be utilised in overcoming the armature volt-drop. $R_a I$.

In a generator $V = E - R_a I$

In a motor $V = E_b + R_a I$ (figure 17.13b and c)

In these figures the induced e.m.f.s E and E_b are drawn in the same direction so that the diagrams refer to the two regimes with the same direction of rotation.

Example 17.3.   A 600-V motor takes 4 A on no-load and 35 A on full load. Its armature resistance is 0.7 Ω. What is the back e.m.f. in each case?

B.E.—8*

No load $R_aI = 0.7 \times 4 = 2.8$ V
∴ $E_b = 600 - 2.8 = 597.2$ V
Full load $R_aI = 0.7 \times 35 = 24.5$ V
∴ $E_b = 600 - 24.5 = 575.5$ V

The Synchronous Machine

The torque production is fundamentally the same as before. With every rotating electrical machine, whatever its type, the m.m.f. distribution of the rotor must rotate at the same speed as that of the stator. The d.c. machine is a special case in which this speed is zero. In the synchronous machine the m.m.f. distribution of the stator, set up by the polyphase currents in its windings, rotates at synchronous speed. The rotor field speed is the same as the rotor speed because of the direct-current excitation. Hence the rotor speed must be the same as the synchronous speed.

Figure 17.14 shows in simplified form the stator and rotor fields. In the alternator the rotor poles are ahead of the stator poles as in figure 17.14a. When the machine is acting as a motor the stator poles are ahead of the rotor poles. In the d.c. motor adjustment to change in load is made by a change in the speed, and therefore in the back e.m.f. For example, in the case of the 600 V motor above, the speed

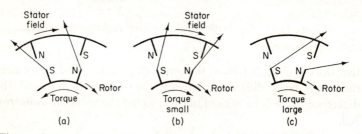

Figure 17.14   Action of the synchronous machine. (a) As generator. (b) and (c) As motor.

when loaded to 3.5 A will be less than for 4 A in order that the difference voltage E_r may be sufficient to admit the necessary current. The synchronous motor cannot change speed, and therefore its adjustment to an increase in load is the falling back of the rotor poles with respect to the stator poles to such an extent that tangential components of the forces of attraction between adjacent stator and rotor poles are sufficient to provide the increased torque. The greater the torque to be developed the greater must be the separation between the adjacent rotor and stator poles in order that the directions of the forces between them may become more and more tangential. With progressively increasing torque a point is reached when the weakening of the force of attraction due to the increased separation predominates. The attractions between the poles are then broken and the motor 'falls out of step'.

Change in the d.c. excitation of the rotor changes the power factor of the motor, and with a sufficiently high value the motor will take a leading current.

The induction motor

The construction of the stator of a three-phase induction motor is like that of a three-phase alternator. The stator winding is connected to the supply and the currents in it produce a rotating field of synchronous speed $N_s = f/p$. The rotor is a laminated cylindrical structure having a winding distributed in slots and, when the motor is in normal operation this winding is short-circuited on itself. There are two types of rotor winding, (a) the phase wound, which is a polyphase winding with its three terminals brought out to step-rings; thus the winding can be connected to external apparatus for starting and regulating purposes. (b) the cage winding which consists simply of round or rectangular bars threaded through the rotor slots and connected together at the two ends by end rings. It follows that the cage winding is permanently short-circuited.

Since the rotor winding is not connected to any supply, either d.c. or a.c., the currents in it must be induced currents. As the only inducing field is the stator rotating field which travels at the speed N_s, and an e.m.f. can only be induced in the rotor when there is relative motion of rotor to stator field, it follows that the rotor can not run at synchronism, but must have a speed somewhat less than this. The difference between n_s and the rotor speed n is called the *slip*. This can be expressed in three ways.

$$N_s - N = \text{absolute slip in rev/s}$$
$$(N_s - N)/N_s = \text{fractional slip, } \sigma$$
$$100 \, (N_s - N)/N_s = \text{percentage slip}$$

Example 17.4. A power station has six-pole synchronous alter-nators running at 1000 rev/min. To its supply is connected a sixteen-pole induction motor whose slip is 2.5 per cent. What is the motor speed?

$$\text{Supply frequency } f = n_\text{s}p = (1000/60) \times 6/2 = 50 \text{ Hz}$$
$$\therefore \text{ Synchronous speed of motor } n_\text{s} = 50/8 = 6.25 \text{ rev/s}$$
$$\text{Slip speed} = 2.5 \text{ per cent of } 6.25$$
$$= 0.156 \text{ rev/s}$$
$$\therefore \quad n = 6.09 \text{ rev/s or } 365.4 \text{ rev/min}$$

An important characteristic of the motor is the frequency of the induced rotor currents. At standstill the stator field sweeps past the rotor conductors at speed n_s. The machine is then, in effect, a

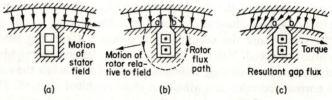

(a)   (b)   (c)

Figure 17.15 Production of torque of an induction motor.

transformer with rotating instead of alternating field. The rotor frequency is therefore the supply frequency f and the fractional slip is unity. At any speed n

$$n_\text{s} - n = \sigma n_\text{s}$$

and therefore at any speed n the rotor frequency is $f_2 = \sigma f$.

Example 17.5. What is the rotor frequency in the case of the motor of example 17.4?

$$\sigma = 2.5/100$$
$$\therefore \quad f_2 = (2.5/100) \times 50 = 1.25 \text{ Hz}$$

This example shows that, during normal operation, the rotor frequency is very low. At the moment of starting it is the supply frequency f, but as the rotor speeds up f_2 progressively diminishes, finally becoming, as shown by the example above, very small.

The manner of torque production is shown by figure 17.15. The stator field is rotating in a clockwise direction and therefore the relative motion of the rotor to the field is counterclockwise, as in the second figure. Applying the screw rule we see that the rotor conductors set up a magnetic flux which, if it existed alone, would take the path shown and would have a counterclockwise direction. Hence, taking both fluxes into account, we see that the air-gap flux is strengthened in the region marked a, and weakened in the region b. The resulting flux distribution is therefore as in the third figure, the lines of force taking an oblique path across the gap. The pull on the rotor is along these lines of force and therefore a torque is set up in a clockwise direction, namely the direction of the stator rotating field.

The figure shows that the stator and rotor m.m.f.s combine to produce a resultant m.m.f. and therefore a resultant field which rotates at synchronous speed.

Load characteristics of electric machines

Figure 17.16 shows diagrammatically three ways in which the connections can be made between armature, field winding and

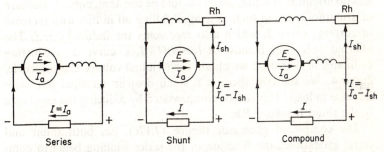

Figure 17.16   Methods of excitation for d.c. machines.

supply line; they are known respectively as series, shunt and compound, and the corresponding curves are shown in figure 17.17. Consider first of all the machines acting as generators. The output current of the series generator (figure 17.17a) is also the field current and therefore the curve of flux per pole will be the magnetisation characteristic, curve 1. Because of the magnetic field set up by the armature currents there is a demagnetising effect which reduces the actual flux per pole to that of curve 2. At constant speed this also

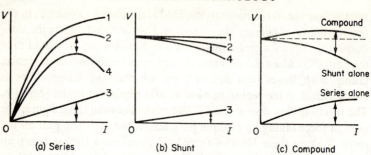

Figure 17.17   Load characteristics of d.c. generators.

gives the induced e.m.f. E. Between the terminals there is the resistance $(R_a + R_{se})$ which produces a volt drop of $(R_a + R_{se})I$, represented by the straight line 3. Deducting the ordinates of curve 3 from curve 2 we obtain the terminal voltage characteristic, curve 4. We see that the series generator is a variable voltage machine. The machine is used very little as a generator.

The excitation of the shunt generator, (figure 17.17b) is from the armature terminals and if the terminal p.d. were constant and there was no armature reaction the flux would be constant, curve 1. Because of armature reaction there is a small falling off in flux with increase in current; curve 2, which also represents the induced e.m.f. The internal volt-drop is now $R_a I_a = R_a(I + I_{sh})$, curve 3. Deducting ordinates of 3 from 2 we obtain the terminal voltage characteristic, curve 4. We see that there is a small drop in terminal p.d. with increase in load. This can be compensated by reducing the resistance in the shunt regulator Rh.

The compound generator (figure 17.17c), has both shunt and series excitations, the function of the series winding being to compensate automatically for the drop due to the shunt winding alone. If there are sufficient series turns the machine can be given a rising characteristic so as to compensate for the volt-drop in the line between generator and load.

Motor characteristics

The back e.m.f., being an induced e.m.f., is given by

$$E_b = \Phi Z n \times P/A$$

Now ZP/A is a constant for a given machine and therefore

$$n \propto E_{\mathrm{b}}/\Phi$$

Again, in a motor

$$E_{\mathrm{b}} = V - RI_{\mathrm{a}}$$

where $R = R_{\mathrm{a}}$ for shunt and $(R_{\mathrm{a}} + R_{\mathrm{se}})$ for series and compound

$$\therefore \ n = k(V - RI_{\mathrm{a}})/\Phi$$

where k is a constant for a given machine. Again RI_{a} is small compared with V so that, approximately

$$n = kV/\Phi$$

and since k and V are constant we have, finally $n \propto 1/\Phi$.

The series machine (figure 17.18) has a small value of Φ at low load and a large value at heavy load, the speed falling off rapidly

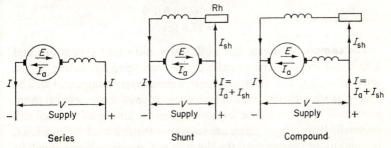

Figure 17.18   Voltage and current variations in d.c. motors.

with increase in load, curve 1. If the load becomes very small the speed will rise to a dangerously high value, and consequently a series motor should never be used in such a way that the load can be thrown off. The most important application is to electric traction, where the throwing off of the load is impossible.

The shunt motor has a constant field current and therefore, but for armature reaction, the flux would be constant, as shown dotted. Actually, the change in flux is very small and we have the expression

$$n = k(V - R_{\mathrm{a}}I_{\mathrm{a}})/\Phi$$

and assuming Φ constant this becomes

$$n \propto V - R_{\mathrm{a}}I_{\mathrm{a}} \propto E_{\mathrm{b}}$$

showing that there is a slight fall in speed with increasing current. If necessary, the speed can be brought back to the no-load value by reducing the shunt current, I_{sh}, that is, by increasing the resistance in the shunt regulator Rh.

Example 17.6.   A 500-V shunt motor takes 3 A when running light, its speed then being 16.7 rev/s. If the shunt resistance is 400 Ω and the armature resistance 1.2 Ω, what will be the speed when the load is such that the intake current is 30 A?

$$\text{(a) No load} \qquad n_0 = 33.3 \text{ rev/s}$$
$$I_{sh} = 500/400 = 1.25 \text{ A}$$
$$\text{armature current } I_{a.0} = 3 - 1.25 = 1.75 \text{ A}$$
$$\therefore \quad E_{b.0} = 500 - 1.75 \times 1.2 = 498 \text{ V}$$
$$\text{(b) On load} \qquad I_a = 30 - 1.25 = 28.75 \text{ A}$$
$$\therefore \quad E_b = 500 - 28.75 \times 112 = 465.5 \text{ V}$$
$$\therefore \quad n/n_0 = E_b/E_{b.0} = 465.5/498 = 0.939$$
$$\therefore \quad n = 16.7 \times 0.939$$
$$= 15.6 \text{ rev/s}$$

The compound motor has both series and shunt excitations and it is usual to have the m.m.f.s of the two windings acting in the same direction. The machine is then cumulatively compounded. The characteristics are intermediate between those of the series and shunt motors. If there are only a few series turns the falling off in speed will be small. If the series turns are such that the full load series m.m.f. is equal to, or greater than the shunt m.m.f. the drop in speed will be very considerable. The characteristic will thus approximate to that of a series motor except that the shunt field will prevent the motor running away on light load or no load.

The torque is proportional to the product ΦI_a. In the series motor at first $\Phi \propto I_a$ so that initially $T \propto I_a^2$. When the flux becomes sensibly constant then $T \propto I_a$ and the characteristic approximates to a straight line through the origin as in figure 17.19a. In a shunt motor Φ is sensibly constant whatever the load and $T \propto I_a$. This gives a straight line through the origin (figure 17.19b). The torque characteristic of the compound motor is also very approximately a straight line through the origin.

The induction motor

This machine is fundamentally a transformer with its primary

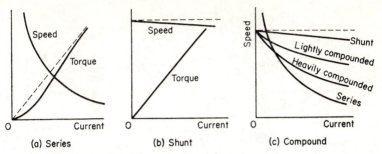

Figure 17.19   Mechanical characteristics of d.c. motors.

winding (the stator winding) supplied at constant p.d. Hence like the transformer it is a constant flux machine. Its speed and torque characteristics are therefore similar to those of the d.c. shunt motor.

D.C. Motor Starters

Since the back e.m.f. is zero at the moment of starting a very heavy current rush will take place if the motor is switched directly on to the line. As a numerical example consider a 440-V compound motor for which the full load line current is 57.5 A, and shunt-current 1 A. The full load armature current is thus 56.5 A. Let $R_a = 0.45$, and $R_{se} = 0.03 \ \Omega$ giving a total resistance in the armature circuit of 0.48 Ω. If this machine is switched directly on to the line the armature current will be

$$I_a = V/I = 440/0.48 = 915 \text{ A}$$

that is, over 16 times the normal current.

Suppose that, for starting purposes the armature current is able to reach 85 A. Then total resistance in the armature circuit must be

$$440/85 = 5.18 \ \Omega$$

But there is already 0.48 Ω in the machine itself and therefore an external resistance of $5.18 - 0.48 = 4.7 \ \Omega$ must be placed in series with the armature circuit. As the armature speeds up the back e.m.f. builds up thus causing the current to fall. The resistance of 4.7 Ω is therefore cut out step by step until finally the motor is directly across the line.

A starter, besides having the necessary resistances and provision for cutting these out step by step, should also possess the following

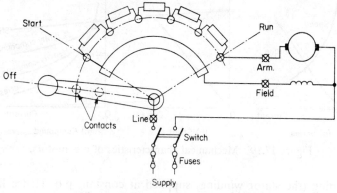

Figure 17.20 Simple motor starter.

automatic features: (a) a device, the no-volt release, which will return the handle to the 'off' position in the event of failure of the supply or opening of the main switch, (b) an automatic overload device with adjustable time delay, and acting independently of the fuses. The scheme of a 'face plate' starter suitable for small motors is shown in figure 17.20.

The no-volt release consists of a small electromagnet energised by the shunt current and arranged to hold the starter handle in the 'on' position against the pull of a spring. If the shunt current ceases then the handle is released and returned to the 'off' position. The overload

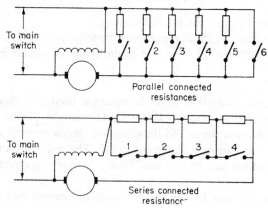

Figure 17.21 Multiple switch starters.

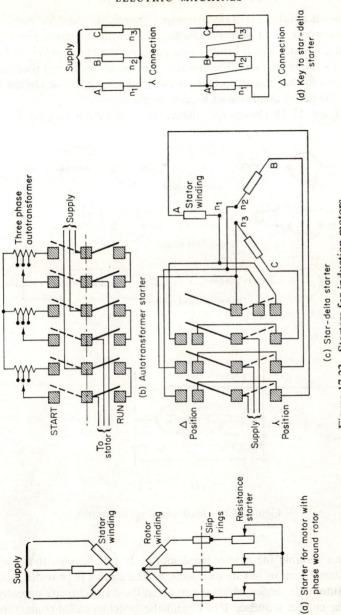

(a) Starter for motor with phase wound rotor

(b) Autotransformer starter

(c) Star-delta starter

(d) Key to star-delta starter

Figure 17.22 Starters for induction motors.

release is another small electromagnet, energised by the line current. In the event of an overload the magnet lifts an armature provided with a contact which bridges two pins and thereby short-circuiting the no-volt release coil. The armature can be set to different positions relative to the magnet so that an overload setting can be chosen to suit the actual condition of loading of the motor.

Figure 17.21 shows the elementary connection scheme for a

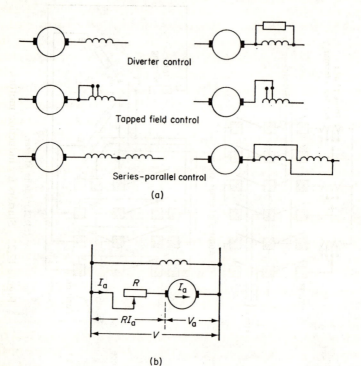

Diverter control

Tapped field control

Series-parallel control

(a)

(b)

Figure 17.23   Methods of speed control.

starter suitable for a very large motor. It will be seen that each resistance section has its own switch. The starter is provided with an automatic device which ensures that these switches are operated in the correct sequence 1 to 6, and the switches, called contactors, are also automatically operated. There are also the usual no-volt and overload releases.

Induction Motor Starters

A motor with a phase-wound rotor can be started by means of a three-phase starting resistance which is cut out step by step as the speed increases (figure 17.22a). The starter can be provided with the same protective devices as the d.c. starter. A small motor with cage rotor can be switched directly on to the line, but with large motors the voltage applied to each phase at starting must be less than the normal voltage. Figure 17.22b shows a starter in which this reduced voltage is supplied by a three-phase autotransformer. It is used to provide a choice of tappings so that the voltage most suited to the starting conditions can be chosen. Figure 17.22c shows a star-delta starter. For normal running the stator phases are connected in delta so that the phase voltage is equal to the line voltage. For starting, the phases are connected in star, the starting voltage per phase thus being $1/\sqrt{3} = 0.577$ of the normal value. It will be seen that the starter is essentially a three-pole double-throw re-grouping switch, with a fourth single-throw blade to provide the neutral point in the star position. The connection scheme should be clear from the key diagram.

Speed control

We have seen that $n \propto E_b/\Phi$ and therefore $\propto V_a/\Phi$ where V_a is the p.d. at the armature terminals. This is an approximate relationship but, as we have seen, V_a is only slightly greater than E_b. Thus the speed can be varied (a) by varying V_a or (b) by varying Φ. The first is called armature control and the second field control. Consider field control first of all. All that is required is the shunt field regulator Rh of figure 17.16b and c. The advantage of the method is that the regulator is a simple appliance and the power loss in it is very small. The disadvantage is that it only gives speeds above the normal, and that with standard industrial type motors the possible speed range is small. It is not possible to vary the flux per pole of a series motor by placing a resistance in series with it because it would be in series with the whole motor and not the field alone. Three possible methods are illustrated in figure 17.23a. In *diverter control* the field current is reduced; in *tapped-field control* the number of field turns is reduced but not the current; in *series-parallel* field control the field current is reduced to one-half.

The simplest method of armature control is rheostatic control by a variable resistance in series with the armature (figure 17.23b).

$$\text{Let } n_0 = \text{speed when } R = 0$$
$$n = \text{speed for any value of } R$$
$$\text{Then } V_a = V - RI_a$$
$$\text{But } n \propto V_a \text{ and } n_0 \propto V$$
$$\therefore \quad \frac{n}{n_0} = \frac{V_a}{V} = \frac{V - RI_a}{V}$$
$$n = n_0 \left(1 - RI_a / V\right)$$

Example 17.7.   A 500-V shunt motor has an armature resistance of 0.24 Ω. The normal full-load current is 82 A and the shunt current 2 A. The no-load speed is 750 rev/min. Draw curves of speed against armature current for the motor running without external resistance, and for controller resistances giving total resistances in the armature circuit of 1, 2, 3, 4, 5 and 6 Ω.

Since the required curves are all straight lines and all start from the point $n = n_0 = 750$ rev/min, it is sufficient to calculate only one more point for each curve, namely that for the current $I = 82$ A

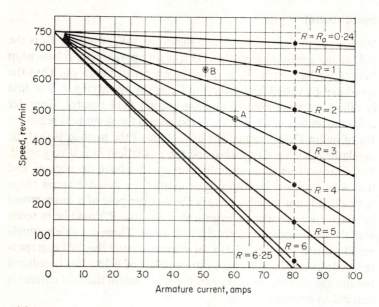

Figure 17.24   Controller characteristics for motor of example 17.7.

$$\therefore \quad I_a = 82 - 2 = 80 \text{ A}$$

(a) no external resistance. $R = R_a = 0.24 \; \Omega$

$$n = 750(1 - 0.24 \times 80/500) = 720 \text{ rev/min}$$

(b) $R = 1 \; \Omega \quad n = 750(1 - 1 \times 80/500) \quad = 630 \text{ rev/min}$

(c) $R = 2 \; \Omega \quad n = 750(1 - 2 \times 80/500) \quad = 510 \text{ rev/min}$

(d) $R = 3 \; \Omega \quad n = 750(1 - 3 \times 80/500) \quad = 390 \text{ rev/min}$

(e) $R = 4 \; \Omega \quad n = 750(1 - 4 \times 80/500) \quad = 270 \text{ rev/min}$

(f) $R = 5 \; \Omega \quad n = 750(1 - 5 \times 80/500) \quad = 150 \text{ rev/min}$

(g) $R = 6 \; \Omega \quad n = 750(1 - 6 \times 80/500) \quad = 30 \text{ rev/min}$

We can also calculate the value of R which will just stall the motor at the above current. For $n = 0$, $RI_a/V = 1$

$$\therefore \quad R = V/I_a = 500/80 = 6.25 \; \Omega$$

The curves for all these resistances are given in figure 17.24. It enables the speed to be read off for any current and any resistance. Thus if the current is 60 A and the resistance 3 Ω, the speed is 480 rev/min as shown by point A. For resistances not represented by any of the graphs, the speed can be estimated. Thus for $I_a = 50$ A and $R = 1.5 \; \Omega$, $n = 637.5$ rev/min (point B).

Rheostatic control is convenient since it provides the whole range from full speed down to zero. The disadvantage is that there is a large power loss, particularly on heavy loads at low speeds. Thus

$$\text{Armature circuit intake} = VI_a$$
$$\text{Armature output} = E_b I_a$$

$\therefore$ Overall efficiency, neglecting shunt-circuit losses.

$$= \text{Output/Intake} = E_b/V = (V - RI_a)/V$$
$$= I - RI_a/V = n/n_0$$

Thus if $n/n_0 = 0.5$ the efficiency is only 0.5. With large motors elaborate methods must be used to dissipate the heat generated in the controller. This illustrates the difference between a starter and a controller. A starter is short-time rated since it carries current for only a short time. A controller is continuously rated.

18 ELECTRONIC DEVICES

The Vacuum Diode

Imagine an enclosure provided with two electrodes, but completely evacuated. If a p.d. is applied no current can flow because of the absence of carriers. If the enclosure contains electrons these will move to the anode, enter it, and become part of an external current, but this current will cease when all the electrons have left the enclosure. For the current to be maintained there must be a continuous supply of electrons and, clearly, these must come from the cathode. The essential condition for a continued flow under steady conditions is that for each electron which enters the anode an electron must be liberated at the cathode. This necessary emission of electrons from the cathode can be accomplished in several ways, and we will consider two of these: field emission and thermionic emission. In the former, electrons at the cathode surface are pulled out by the attraction of an intense electric field, and we shall see that this necessitates positive as well as negative carriers and therefore does not apply when the carriers are electrons only. In the latter the cathode temperature is raised. This causes the ions of the metal to oscillate and collisions between ions and electrons at the surface will, if sufficiently violent, cause electron emission. Clearly the electron emission will increase as the temperature is increased.

For adequate emission from hot cathodes pure metals, such as platinum or nickel must be operated at high temperature, with platinum it can be as high as 2500 K. By coating a metallic cathode with oxides of barium, calcium or strontium, emission can be obtained at very much lower temperatures, say up to 1200 K.

Space charge

If no p.d. is applied to the electrodes, the electrons from a hot cathode will accumulate within the enclosure, and owing to their mutual repulsions many will return to the cathode. A state of equilibrium will be reached when the number so returned is equal to the number emitted. This accumulation of electrons is called a negative space charge; it has the greatest density adjacent to the cathode, as indicated in figure 18.1. Now let the anode be made positive with respect to the cathode, by connection to an external source, say a battery as in figure 18.2, then electrons can escape into the external

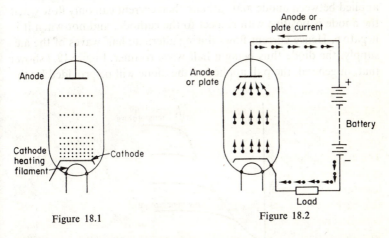

Figure 18.1

Figure 18.2

Figure 18.1   Negative space charge.
Figure 18.2   Hot cathode diode on load. Source for heating filament not shown.

circuit via the anode, and current will flow. If the battery voltage is increased the current will increase until the rate at which electrons are emitted at the anode is equal to the rate at which they enter the anode. For a given rate of emission the maximum current is now obtained and further increase in anode voltage has little effect on the current. The diode is then said to be saturated, (figure 18.3). In order to increase the current the rate of emission of electrons from the cathode must be increased, and therefore the cathode temperature must be increased. Thus a diode can be given any desired saturation current, within the limits of its operation, by a suitable adjustment of the cathode temperature. Because of the effect of the space charge, the vacuum diode is said to be space charge controlled. In figures 18.1 and 2 the source for the current to the heater is not shown.

The Vacuum-type Diode Rectifier

Since electrons can only pass into the external circuit via the anode it follows that current can pass through the diode in one direction only, namely from anode to cathode in the conventional sense. The diode can therefore be used as a rectifier. If an alternating p.d. is applied between anode and cathode then current can only flow when the anode is positive with respect to the cathode, and not when it is negative. Hence current flows during alternate half waves of the a.c. supply, the diode thus being a half-wave rectifier. Figure 18.4 shows that, in general, the p.d. applied to the diode will be provided by the

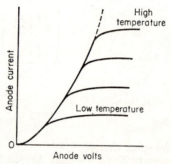

Figure 18.3   Dependence of diode output on cathode temperature.

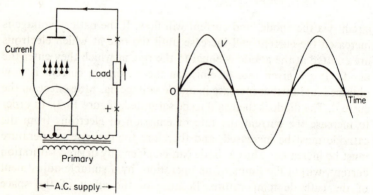

Figure 18.4   Hot cathode diode used as a half-wave rectifier.

secondary winding of a transformer, which will also have a small third or 'tertiary' winding for the supply to the cathode, which is shown directly heated for simplicity. The centre point of this winding is taken to the main secondary winding, as shown. Such a connection is called a centre-tap.

The mean voltage of an alternating wave is $(2/\pi)V_{\mathrm{m}}$ and therefore

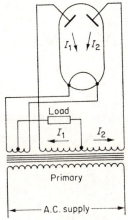

Figure 18.5   Single-phase hot cathode full-wave rectifier.

the mean over the whole periodic time when only one half cycle is utilised is V_{m}/π. If the mean value of the rectified current is I_{av} then, for the same reason, its peak value is πI_{av}.

Power during conducting half period $\quad I^2 R = \frac{1}{2}I_{\mathrm{m}}^2 R$
$$= \frac{1}{2}\pi^2 I_{\mathrm{av}}^2 R$$
Power during non-conducting half period $\quad = 0$
∴ Average power during whole period $\quad = \frac{1}{4}\pi^2 I_{\mathrm{av}}^2 R$
∴ r.m.s. current $\quad = (\frac{1}{4}\pi^2 I_{\mathrm{av}}^2)^{\frac{1}{2}}$
$$= 1.57\, I_{\mathrm{av}}$$

During the non-conducting half-period the anode voltage is reversed and acquires a maximum value of $-V_{\mathrm{m}}$ relative to the cathode. Hence the diode must be able to withstand a peak inverse voltage of V_{m}.

Because of the intermittent nature of its current output and its poor utilisation of the transformer the half-wave rectifier is little used. By providing the diode with two anodes, it becomes a full-wave rectifier, figure 18.5. One end of the load is connected to the centre point of the transformer so that the rectifier is, in effect, two half-wave rectifiers in the same enclosure, one operating on forward voltage, the other on reverse voltage. The rectified current wave is now a series of half sinusoids, as in figure 18.6. For a given value

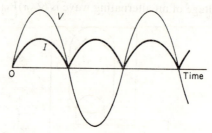

Figure 18.6   Full-wave rectification.

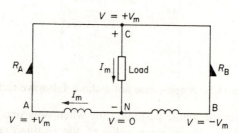

Figure 18.7   Illustrating the peak reverse voltage.

of I_{av} it is clear that with full-wave rectification I_m is one-half of that with half-wave.

$$\therefore \ I_m = (\pi/2)I_{av}$$

The power during one conducting period is

$$P = \tfrac{1}{2}I_m^2 R = \tfrac{1}{2} \times \frac{\pi^2}{4} I_{av}^2 R$$

$$\therefore \ I_{r.m.s.} = (\tfrac{1}{2} \times \frac{\pi^2}{4} I_{av}^2)^{\frac{1}{2}} = 1.11 I_{av}$$

as against $1.57\,I_{av}$ for half-wave rectification.

Now consider the skeleton diagram of figure 18.7 in which the two halves of the double diode are represented conventionally by the arrows. Consider the instant the e.m.f. in the half transformer NA is a maximum, V_m. Then if we regard the point N as at zero potential, A will be at $+V_m$ and B at $-V_m$. Neglecting the volt-drop in conducting rectifier R_A the potential of point C will also be $+V_m$. Hence the non-conducting rectifier R_B will be subjected to a p.d. of $2V_m$. In practice it will be somewhat less than this because the rectifier volt-drop is not zero. The full-wave rectifier must therefore be able to withstand a peak reverse voltage of $2V_m$.

The Gas-filled Diode

Suppose that there is a gas or vapour at low pressure within the enclosure. The electrons, owing to their minute mass will acquire very high velocities under the influence of the potential gradient, and therefore high kinetic energies, $\frac{1}{2}mv^2$. They will collide with the

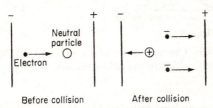

Figure 18.8   Collision of an electron with a neutral particle.

neutral particles and, if the collision is sufficiently violent, these particles will lose an electron and become positive ions. Thus, starting with one negative carrier and one neutral particle there results from the ionising collision, two negative carriers and one positive carrier, figure 18.8. Each electron will again take part in collision processes so that the process becomes cumulative, resulting in many electrons and many positive ions. The process is called an 'electron avalanche'. Because of the positive ions the gas-filled diode is no longer space-charge controlled. Its resistance is very low and its current-voltage characteristic is almost a vertical straight line, figure 18.9. It can therefore handle large powers whereas the vacuum diode, whose current-carrying capacity is dependent on electron emission only, is of small output and its resistance high.

The cold-cathode rectifier

With a relatively cold cathode the contribution to the total emission by thermionic emission is small. Most of it is due to field emission and secondary emission. Imagine vast numbers of positive ions approaching the cathode. When a 'front' of ions is very close to the cathode an electrical double layer will result, figure 18.10, in which there will be an intense electric field. The attraction of electrons at the surface by this field will result in their escape from the cathode.

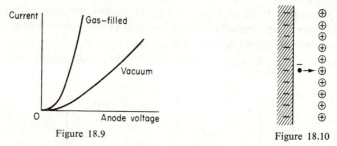

Figure 18.9

Figure 18.10

Figure 18.9   Characteristics of vacuum and gas-filled diodes.
Figure 18.10   Illustrating field emission.

In the case of secondary emission ions bombard the cathode surface and so detach electrons. This type of emission is most effective with very heavy ions like those of mercury, and it is therefore of importance in the operation of the mercury-arc rectifier, in which the cathode is in the form of a pool of liquid mercury. In this rectifier the current through the enclosure is in the form of an arc whose termination at the mercury pool is a very small area called the cathode spot. This spot moves erratically over the surface under the influence of electric and thermal forces; it is an area of high temperature and therefore contributes thermionic emission to the total emission.

Reservoir Capacitors

This is a capacitor connected in parallel with the load in order to smooth the rectifier output. The capacitor becomes charged during the conducting half-period and discharges into the load during the non-conducting half-period. This is possible because of the reverse characteristic of the rectifier as shown in figure 18.11, in which a half-wave rectifier is shown. It is assumed that it is an 'ideal' rectifier, that is, one in which the reverse resistance is infinite and the reverse current therefore zero. Figure 18.11c shows that the capacitor,

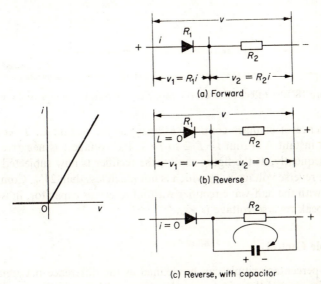

(a) Forward

(b) Reverse

(c) Reverse, with capacitor

Figure 18.11 Effect of capacitator on performance of half-wave rectifier.

charged to the polarity shown during the conducting half-period, will discharge in the right direction through R_2 during the non-conducting half-period, thereby maintaining the flow of current.

The current i can only flow when v is greater than v_2. It will therefore commence at some point A in the wave of v, shown in figure 18.12. Part of i flows through R as the load current i_2; the difference $(i - i_2)$ charges the capacitor. At some instant B at which v ceases to be greater than v_2 the rectifier ceases to conduct and i becomes zero. Hence the current i taken from the a.c. supply, or the secondary of the transformer, is in the form of a series of pulses, as shown. The moment i becomes zero, the capacitor begins to discharge into the resistor R_2, and we see from figure 18.11c that the capacitor load circuit is a RC series circuit the current therefore falling at a rate determined by the relative values of R and C. This

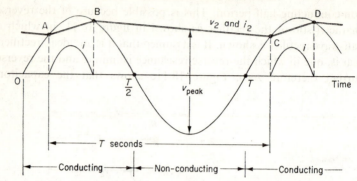

Figure 18.12   Effect of reservoir capacitor on output voltage and current.

sequence of changes commences again at the instant C, T seconds after instant A. From $t=T/2$ to $t=T$, the voltage v is negative and consequently, as the figure shows, the rectifier is now subjected to a peak reverse voltage v_{peak} which is not much less than $2V_m$. Compare this with the half-wave rectifier without reservoir capacitor, in which the peak reverse voltage is V_m.

Ripple Control

The percentage ripple can be defined as the difference between the maximum and the minimum values expressed as a percentage of the maximum value. The percentage ripple of the single-phase rectifiers without smoothing apparatus such as a reservoir capacitor is therefore 100 per cent. The reservoir capacitor is therefore widely used for small outputs but it has two disadvantages, (a) the high peak reverse voltage as explained above, (b) the impulsive nature of the current from the supply transformer. Since the peak value of this current may be several times the output current, the utilisation of the supply transformer is very poor.

    A simple method, possessing none of the above disadvantages, is to connect an inductor in series with the load. Since inductive reactance is proportional to the frequency, the current passed by an inductor is inversely proportional to the frequency. Any non-sinusoidal, but periodic, wave can be resolved into a series of sinusoidal components of frequencies f, $2f$, $3f$, and so on, the presence of even harmonics depending on whether or not the negative half-wave is a repetition of the positive half. As a purely illustrative example suppose that a p.d. for which $V_{1m} = 100$, $V_{2m} = 40$ and $V_{2m} = 20$ is applied to an inductor of reactance 10 Ω at the fundamental frequency, then

$$I_{1.m} = 100/10 \qquad = 10\ \text{A} = 100 \text{ per cent say}$$
$$I_{2.m} = \quad 40/10 \times 2 = \quad 2\ \text{A} = 20 \text{ per cent}$$
$$I_{3.m} = \quad 20/10 \times 3 = 0.67 \quad = 6.7 \text{ per cent}$$

We see that the harmonics in the current wave are of smaller amplitude than the corresponding harmonics in the voltage wave. If the

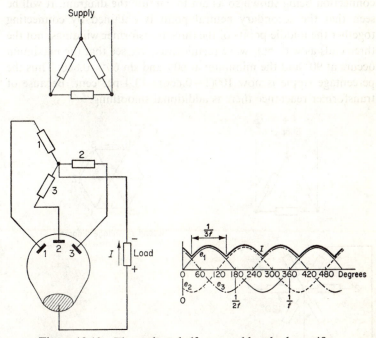

Figure 18.13   Three-phase half-wave cold-cathode rectifier.

load is a resistor and a smoothing inductor is connected in series with it then, although smoothing will take place, it will be to a much smaller extent.

The best way to reduce ripple is to use a rectifier for which the ripple in the output voltage is small, that is, without the addition of external smoothing devices. This can be done by increasing the number of phases. Figure 18.13 shows the connection scheme for a three-phase half-wave rectifier with a resistance load, so that the current waveform is the same as the voltage waveform. The rectifier is half-wave because the negative half-waves are not utilised. We see that the maximum current occurs at 90° and the minimum at 30°, and since sin 30° = 0.5 we see that the theoretical percentage ripple is 50 per cent. The arc stays on anode 1 so long as its voltage e_1 is greater than e_2. When $e_1 = e_2$, then e_2 is about to change from anode 1 to anode 2, and so on. Thus the arc commutates from one anode to another.

Figure 18.14 shows a three-phase full-wave rectifier, only one anode connection being shown so as not to confuse the diagram. It will be seen that the secondary neutral point is obtained by connecting together the middle points of the three transformer windings, not the three ends as with half-wave rectification. We see that the maximum occurs at 90° and the minimum at 60°, and sin 60° = 0.866. Thus the percentage ripple is now $100(1 - 0.866) = 13.4$ per cent. Because of transformer reactance there is additional smoothing.

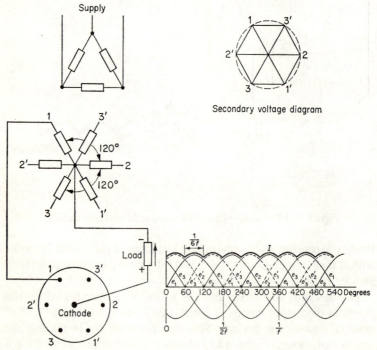

Figure 18.14   Three-phase full-wave rectifier.

It will be seen from the wave diagram that each anode in turn delivers an almost rectangular block of current, that is, an almost constant direct-current. With the simple connections shown the transformer secondaries thus carry direct current and this has the effect of magnetically saturating the transformer cores. To avoid this much more complex circuits must be used, but these are beyond the scope of this book.

Metal Rectifiers

If current is passed through a junction of two dissimilar conductors it is found that, in some cases, resistance to current flow in one

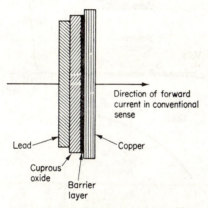

Figure 18.15   Copper oxide rectifier.

direction is very high, while to flow in the opposite direction it is low. If two conductors can be found in which the resistance to one direction is exceedingly high then clearly, the junction will act as a rectifier with a small reverse current. For example, the boundary between cuprous oxide and copper. Cuprous oxide is formed py heating pure copper in air to about 1300 K, and the copper-oxide rectifier consists essentially of a disc of copper with a layer of cuprous oxide on one side. To facilitate electrical contact, the oxide layer has sputtered on to it a thin metallic coating, this coating forming one contact and the copper disc the other. Current passes easily from oxide to copper. Contact with the metallic coating of the oxide is made by a plate of lead, the construction being as shown diagrammatically in figure 18.15.

The forward current is of the order of one-third of an ampere for a disc of 2 cm diameter, while the maximum safe reverse voltage is 12 V. The characteristics at 20° C for a single unit are given in figure 18.16, from which it will be seen that the current-voltage relationship does not follow a straight line law, that is, is not linear. For operation at a voltage higher than is suitable for a single unit, an appropriate number are threaded on to an insulating rod, and

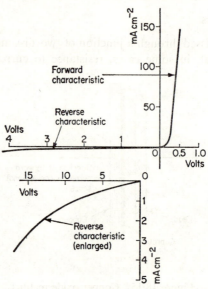

Figure 18.16 Characteristics of copper oxide rectifier at 20° C.

contact between adjacent units made by means of lead washers. At every two or three units a larger diameter metal fin is added so as to facilitate cooling.

Thus the rectifier is a diode consisting of a junction between cuprous oxide, which is a semiconductor (see p. 254), and a metal. The semiconductor possesses few free electrons whereas the metal possesses the normal number. Very simply, the state of affairs is that of figure 18.17. If a p.d. is applied in such a direction as to make the metal positive with respect to the oxide (figure 18.17a), the effect will be to prevent any of the electrons in the metal from passing into the oxide. The electrons will be repelled from the junction thereby forming a barrier layer. If the oxide is made positive with respect to

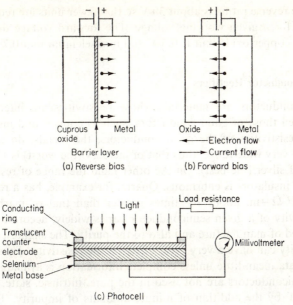

Cuprous oxide   Metal

Barrier layer

(a) Reverse bias

Oxide   Metal

— Electron flow
→ Current flow

(b) Forward bias

Conducting ring
Translucent counter electrode
Selenium
Metal base

Light

Load resistance

Millivoltmeter

(c) Photocell

Figure 18.17 Metal rectifiers.

the metal, as in figure 18.17b, then electrons in the metal will be attracted across the junction, which no longer acts as a barrier.

The selenium rectifier is also a diode consisting of a junction between a disc of selenium, which like cuprous oxide is deficient in electrons, and a metal. The action thus corresponds to figure 18.17, a barrier layer being formed if the selenium is negative with respect to the metal. The construction is as follows: molten selenium is deposited on a nickel-plated iron base plate and a translucent electrode of low-melting-point alloy such as Woods metal, is deposited on the selenium surface. The insulating barrier between the selenium and the alloy 'counter-electrode' forms spontaneously. The efficiency of the contact can be increased by passing a small rectified alternating current through the cell for several hours in the direction of high resistance. Since the 'front' of the cell is a trans-lucent layer it admits radiant energy, such as light. Consequently if the rectifier is connected to a millivoltmeter as in figure 18,17c and the front of the cell irradiated, the voltmeter will be deflected. Using a suitable series resistance the cell will give a deflection directly proportional to the intensity of the radiation.

The reverse rating is about 30 V so that fewer units are required in series for a given working voltage. The forward voltage drop in a single copper oxide unit is 0.3 V and in a selenium unit 0.7 V.

Semiconductor Rectifiers

Semiconductors are materials whose resistivities are intermediate between those of good conductors, such as copper, and insulators. The resistivities of the good conductors, the metals, do not vary over a very wide range, thus that of mercury (the worst) is 100 times that of silver (the best). On the other hand the range of resistivities in the insulators is enormous. Quartz, for example, has a resistivity of $10^{17} \ \Omega-m$, about 10^9 times greater than that of marble. The resistivity of a given semiconductor varies widely, according to the method of manufacture and also to the purity. The slightest trace of impurity can have a very great effect on the resistivity. Thus resistivity data mean little unless complete information is available.

Semiconductors are not used in the pure, intrinsic, state, but are 'doped' by the addition of minute amounts of impurity. They are then called extrinsic semiconductors. The impurity to be used is decided by the required properties of the extrinsic semiconductor.

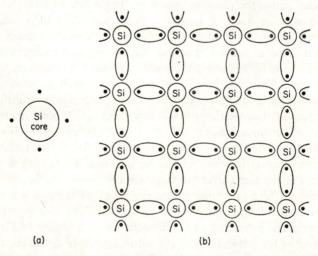

(a)

(b)

Figure 18.18.   (a) Silicon atom. (b) Structure of intrinsic silicon showing how neighbouring atoms are held together by a system of covalent bonds.

The action of an impurity is as follows: we will consider the case of silicon only, that of germanium being very similar. Silicon has a tetravalent atom, that is, the core, consisting of the nucleus and the completed shells has external to it four valency electrons, as in figure 18.18a. The crystal of silicon consists of silicon atoms arranged in a definite geometrical pattern in which adjacent atoms are held together by means of a covalent bond. The structure is, of course, three-dimensional but it is shown flat in figure 18.18b. If an atom is removed, then another atom can take its place if it also has four valency electrons. If it has more than, or less than, four then an adjustment must take place and it is this adjustment which gives the extrinsic semiconductor its special properties.

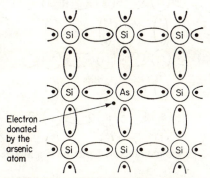

Figure 18.19   n-type semiconductor.

Suppose first of all that the impurity activator is arsenic. This has five valency electrons and therefore, in order that it may fit into the crystal lattice of tetravalent silicon it must shed one electron as shown in figure 18.19. The arsenic is therefore called a donor impurity. Since, in the spaces between the atoms, there are as many additional electrons as there are impurity atoms, the semiconductor possesses more free electrons than the intrinsic form. It therefore has negative charges in addition to the free electrons of the instrinsic material. These free electrons are called thermal electrons because their motion is largely determined by the temperature. The semi-conductor is therefore known as the n-type. Thus donor impurities produce n-type semiconductors.

Now suppose that the impurity activator is indium which is an element whose atom has only three valency electrons. In order to fit into the crystal lattice it must acquire a fourth electron, and as the process is essentially one of crystal structure re-adjustment, this additional electron must be one of the electrons in the lattice and not a free electron. In order to accomplish this one of the covalent bonds must be broken as shown in figure 18.20. Now a substitution in the crystal lattice of an atom with only three electrons in the place of one with the normal four electrons means that there is an electron short. If a negative charge is removed from a previous stable assembly of positive and negative charge then, relatively, the gap left by the absence of this charge acquires positive electrification, and

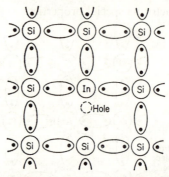

Figure 18.20   p-type semiconductor. Note the incomplete valency bonds.

is therefore called a positive hole. It is represented by the small dashed circle in the figure. The semiconductor now possesses positive carriers and is called 'p-type'. The indium is called an acceptor. Thus acceptor impurities produce p-type semiconductors.

The hole does not stay in one place but is mobile although its mode of progression is very different from that of a free electron. During its travels, no matter how complex, an electron retains its identity; for example, if an electron moves from A to B then the electron which arrives at B is the one which started at A. This is not the case with holes. The process of hole mobility is complex and perhaps the simplest thing is to say that the removal of a charged particle from a previously stable system sets up instability. This in turn results in an attempt at adjustment, this involving motion, but again, it must be associated with the lattice structure. The holes therefore move from one covalent bond to another, and although the movement of an individual hole may be random, that of the holes *in toto* is to ensure a uniform distribution throughout the structure. A hole, being positive, is able to attract an electron from a neighbouring complete covalent bond, with the result that it becomes neutralised and a new hole, in another place, is created. This new hole, in its turn, becomes neutralised and produces another hole; and so the process continues. Thus hole mobility is a process of neutralisation of an existing hole and the creation of a new hole in a different place. The mobility of holes is about one-half that of free electrons. The process is illustrated in figure 18.21.

Operation of semiconductor diodes

During the manufacture of a semiconductor diode regions of both p-type and n-type are formed in the same piece of originally intrinsic semiconductor. The junction is therefore one of intimate association and not merely a butt contact between the two types.

Figure 18.22 shows in a simplified manner the operation of the diode. In figure 18.22a no p.d. is applied and therefore there is no bias. Because of the superior mobility of electrons to holes there will

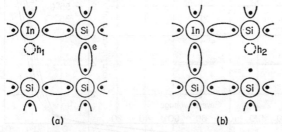

Figure 18.21 Hole mobility. The electron e (a) is attracted by the hole h_1 which it neutralises and at the same time creates another hole, h_2. The process is thus equivalent to the movement charge $+e$ from h_1 to h_2.

be a net migration of electrons from the n-type to the p-type and, as a result, the n-type will become less negative and the p-type more positive so that, in effect, the junction will become a potential barrier with the side adjacent to the n-type positive with respect to the side adjacent to the p-type. Since the barrier is formed by the removal of electrons from the n-type, it is also called a depletion layer. If the diode is reverse biased as in figure 18.22b the p.d. across the depletion layer is increased, the thickness of the layer is increased and the

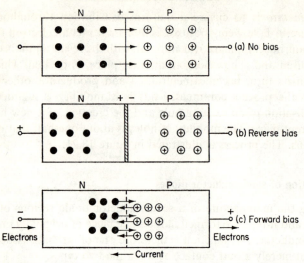

Figure 18.22   Operation of the semiconductor diode.

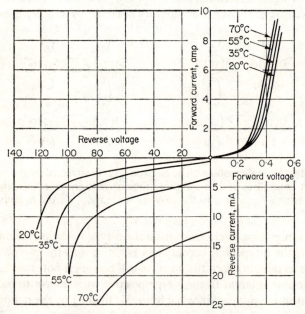

Figure 18.23   Static characteristics of the EW54 diode at various ambient temperatures, showing the temperature dependence of the characteristics. (*Courtesy of General Electric Co., Ltd.*)

mobile carriers on both sides are expelled away from the layer. There will, however, be a small reverse current because of the thermally produced electrons in both n- and p-types. If the diode is forward biased, the p.d. across the junction is neutralised, and both electrons and holes move towards the junction and are able to cross it. The potential barrier is thus broken down and forward current can flow.

Figure 18.23 gives the characteristics of one model of germanium diode. It will be seen that the performance is largely dependent on temperature. The reverse characteristic tends to become vertical as the reverse voltage increases. At a value called the Zener voltage the potential gradient at the barrier becomes sufficient to increase the electron flow across it to such an extent that the reverse characteristic becomes almost vertical, but this does not necessarily result in a breakdown of the diode. Provided that the diode is designed to operate at the Zener level it can be used as a voltage regulator because the voltage across the junction remains constant over a wide range of reverse current.

Breakdown of Diodes

1. A high reverse voltage can impart so much energy to the mobile electrons that these can liberate by collision with atoms, other electrons. This process is cumulative and is similar to the avalanche phenomena in gases. The reverse resistance is very high and therefore if the avalanche process results in an I^2R loss so high that the rate of heat production is greater than the rate of heat dissipation, thermal runaway, or thermal instability, will result, and the diode will be destroyed.

2. In Zener breakdown a high electric field across the depletion layer detaches electrons initially bound to the rest of the atom, thereby leaving holes and creating electron-hole pairs and therefore two carriers for each electron detached. This process results in increased reverse current and therefore increased I^2R loss.

Direct 'punch through', that is, the sudden breakdown due to the destruction of the insulating properties by a high electric field is unlikely because the design is such that one of the above processes will take place before this can occur.

It follows that the successful operation of a semiconductor diode is dependent on efficient heat dissipation. With cuprous oxide and selenium rectifiers it is sufficient to provide cooling fins between every three or four units. With the germanium and silicon diodes used for large power rectifications, the diodes are mounted on a heat-sink in the form of an extruded metal bar whose cross-section gives the maximum dissipating surface, (figure 18.24). The heat dissipation can be increased, if necessary, by blowing cold air along the fins or along the heat sink.

The construction of the silicon power rectifying diode is shown in figure 18.24. It will be seen that the diode is totally enclosed or encapsulated. This is because most of the reverse current flows

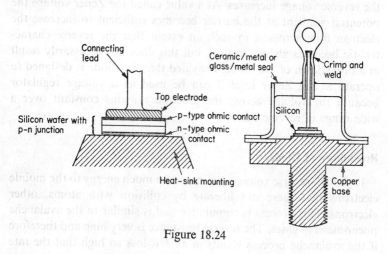

Figure 18.24

through the surface layer which is extremely sensitive to contamination. Before encapsulation all contaminants have to be removed from the surface of the rectifier.

Comparison of the various types

The reverse rating of the cuprous oxide rectifier is 6 to 12 V, and that of the selenium rectifier 30 to 40 V. The forward voltage drops are 0.5 V and 0.7 V and the operating temperatures about 300 K and 310 K respectively. The temperatures at which thermal instability begins are 330 K and 430 K respectively. These rectifiers are therefore only suitable for low powers; they are widely used in its instrumentation.

With germanium and silicon rectifiers the peak reverse ratings are 300 V and ? V and the forward voltage drops 0.5 V and 1.1 V respectively. Germanium has a low temperature for thermal instability and if this is exceeded the diode fails instantly. Even with efficient cooling, the junction temperature should not exceed 370 K, the usual figure being 350 K. Silicon is better in this respect since the junction temperature can be as high as 470 K. Instantaneous failure occurs if the temperature exceeds 470 K.

Another useful basis of comparison is the output per unit volume of rectifier material. Calling that for the cupric-oxide rectifier 1, the values for the others are: selenium 12, germanium 36, silicon 100. The above data shows that silicon is the most important of the semiconductor rectifiers, in fact it is rapidly replacing the mercury-arc rectifier even for the highest powers.

Rectifier Connections

Figure 18.25 shows a number of commonly used rectifier circuits. Figure 18.25a and b are simply half-wave and full-wave arrangements

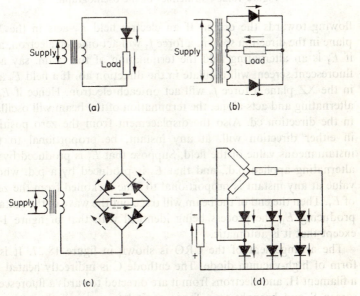

Figure 18.25 Semiconductor rectifier connections.

respectively. Figure 18.25c is a very commonly used bridge circuit but limited to semiconductor rectifiers. Figure 18.25d is a three-phase bridge circuit again only used with semiconductor rectifiers. The reason is, that with hot cathode gas-filled diodes, there would be too much complications and expense because of cathode heating apparatus.

The Cathode-ray Oscilloscope (CRO)

Figure 18.26 shows a beam of electrons projected in the OX direction from a cathode. It is equivalent to a current in a flexible conductor

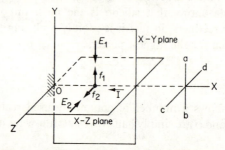

Figure 18.26　Principles of the oscillograph.

flowing towards the origin. If an electric field E_1 acts in the XY plane in the direction shown, a force f_1 will act on each electron, and if E_1 is an alternating field the termination of the beam, say at a fluorescent screen, will oscillate in the direction ab. If a field E_2 acts in the XZ plane a force f_2 will act on each electron. Hence if E_2 is alternating and acts alone, the termination of the beam will oscillate in the direction cd. Also the displacement from the zero position in either direction will, at any instant, be proportional to the instantaneous value of the field. Suppose that E_1 is produced by an alternating applied p.d., and that E_2 is produced by a p.d. whose value at any instant is proportional to time reckoned from the zero of E_1. Then the end of the beam will trace out the wave of the voltage producing E_1, the process being identical with that of figure 14.1 except that it is automatic.

The arrangement of the CRO is shown in figure 18.27. It is a form of high-vacuum diode. The cathode C is indirectly heated by a filament H, and electrons from it are directed towards a fluorescent screen S and brought to a focus on it by means of the array of electrodes shown. The first is the grid G whose potential with respect to C can be adjusted so as to vary the electron current. The second is an accelerating electrode K, which imparts a high velocity to the electrons in the axial direction. This is at a high potential often several thousand volts above C. The third J, is a focussing electrode whose potential is adjusted to bring the spot of light on the screen to a sharp focus. The fourth A is the anode, operated at a high voltage with respect to C so as to give further acceleration to the electrons. This assembly of electrodes is usually called the 'gun'.

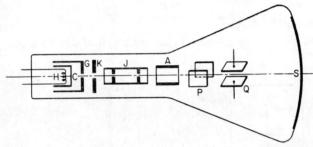

Figure 18.27   Arrangement of the electrode and deflecting plates in a cathode-ray oscilloscope.

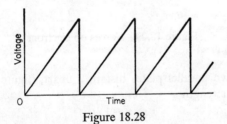

Figure 18.28

After leaving A the electron beam passes, in turn, between two pairs of deflecting plates P and Q, one pair vertical, the other horizontal. The p.d. giving the field E_1 is applied to the horizontal plates Q, and the p.d. giving the time to plates P. The magnitude of E_2 must be proportional to the time reckoned from the zero of the wave to be traced, and after the completion of the wave it must return to zero. The sweep voltage is therefore of saw-tooth shape, as shown in figure 18.28. The manner in which this particular voltage variation is obtained is beyond the scope of this book.

Since the only carriers of electricity inside the tube are electrons it follows that the operation is dependent on the forces which are made to act on the electrons. These are the accelerating forces produced by the gun to give the electrons a high velocity by the time they reach the deflecting plates and the forces due to the potential differences applied to the plates. When deflection is due to one pair, or both pairs of plates, the forces are electrostatic. Deflection can also be produced electromagnetically by means of a pair of current-carrying coils arranged one on each side of the tube in the region of the deflecting plates and with their common axis at right-angles to the tube axis.

Electrostatic deflection

Let an electron be situated in an electric field due to a p.d. of V

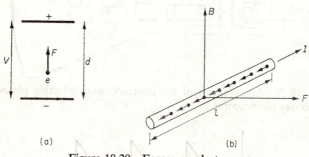

Figure 18.29 Forces on electrons.

applied to two parallel plates distant d apart, (figure 18.29). Then electric field strength

$$E = V/d$$

Force on the electron $F = Ee$ acting towards the positive plate.
Acceleration of electron $a = F/m$ acting in the same direction as F.

Example 18.1. Two parallel plates 5 cm apart are connected to a 500 V, a.c. supply. Assuming that an electron starts from the negative plate calculate (a) the force on the electron, (b) its acceleration, (c) its velocity by the time it reaches the positive plate, (d) the time taken to reach this plate.

Take $e = 1.6 \times 10^{-19}$C and $m = 9.1 \times 10^{-31}$kg.

$$E = 500/5 = 100 \text{ V/cm} = 10^4 \text{ V/m}$$

$$F = Ee = 10^4 \times 1.6 \times 10^{-19} = 1.6 \times 10^{-15} \text{ N}$$
$$a = F/m = (1.6 \times 10^{-15})/(9.1 \times 10^{-31}) = 1.76 \times 10^{15} \text{ m/s}$$
$$d = \tfrac{1}{2}at^2$$
$$\therefore \quad t = (2d/a)^{\frac{1}{2}} = (2 \times 5 \times 10^{-2}/1.76 \times 10^{15})^{\frac{1}{2}}$$
$$= 7.5 \times 10^{-9} \text{ s}$$
$$v = at, \text{ since the initial velocity is zero}$$
$$= 1.76 \times 10^{15} \times 7.5 \times 10^{-9}$$
$$= 1.32 \times 10^7 \text{ m/s}$$

The positive energy acquired by the electron

$$W = eV = 1.6 \times 10^{-19} \times 500 = 8 \times 10^{-17} \text{ J}$$
$$\text{or } W = Fd = 1.6 \times 10^{-15} \times 5 \times 10^{-2} = 8 \times 10^{-17} \text{ J}$$

Electromagnetic deflection

A magnetic field can only produce a force on an electron when the
electron is in motion. Consider a parallel pencil of electrons having
a uniform distribution over the cross-section of the pencil. Let the
velocity be v, then each electron carries electricity at the rate of ev
coulombs per second. Suppose that in unit length there are n
electrons, then, across any normal cross-section the moving electrons
will transfer nev coulombs per second. Now the ampere corresponds
to an axial movement of one coulomb per second, showing that the
electron pencil is equivalent to a current of

$$I = nev \text{ amperes}$$

If this pencil, of length l, traverses a uniform magnetic field of
intensity B tesla in a direction at right-angles to the direction of the
magnetic lines of force, as in figure 18.29b there will be exerted a
force F of magnitude

$$F = BIl \text{ newtons}$$
$$= Bnevl \text{ newtons}$$

In this length l these will be nl electrons so that force on one electron

$$F = Bev \text{ newtons}$$

and it will produce an acceleration in the direction of F, of

$$a = F/m = Bev/m \text{ m/s}^2$$

Example 18.2. An electron has a velocity of 2×10^7 m/s when
it enters, perpendicularly to the lines of force, a magnetic field

of strength 5 T. What is the magnetic force on the electron?

$$F = Bev$$
$$= 5 \times 1.6 \times 10^{-19} \times 2 \times 10^7 = 1.6 \times 10^{-11} \text{ N}$$

Focussing of the electron beam

If an uncontrolled beam of electrons leaves a cathode then, because of the repulsions of the electron for one another, the beam will spread out and will not focus on the CRO screen. It can be focussed either electrostatically or magnetically and we will consider the electrostatic case only. It was stated that focussing was accomplished by the tube J, figure 18.27. Figure 18.30 shows a cathode C surrounded

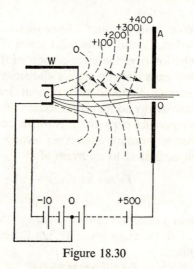

Figure 18.30

by such a tube W. The tube is negative with respect to the cathode, its potential being −10 V. The anode A is at a potential of 500 V. The electric field set up by such an arrangement is such that equi-potential planes have the shape shown dotted. Since an electron moves from a point of low to one of higher potential, the forces on the electrons emitted by C will be at right-angles to the dotted equipotential lines, and therefore in the direction of the arrows. These all point inwards and thus prevent any outwards spread of the electron beam. The paths of the electrons are thus as indicated by the full lines in the lower half of the diagram.

19 ELECTRICAL MEASURING INSTRUMENTS

Indicating Instruments, Ammeters, Voltmeters, Wattmeters

There are three essential features:

(i) A device to produce a deflecting torque on the moving part of the instrument.

(ii) A controlling device which exerts a torque in opposition to the deflecting torque whenever a deflection takes place. The deflection is that which makes these two torques equal.

(iii) A damping device to give an aperiodic motion to the moving part; that is a motion such that, if the indicating needle overshoots the mark it is brought to the correct position without oscillation.

Figure 19.1 shows some controlling and damping devices; the

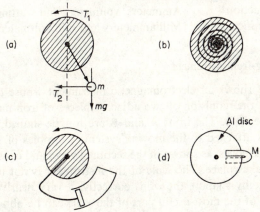

Figure 19.1 Controlling and damping devices.

moving system is shown purely diagrammatically as a shaded circle, figure 19.1a–c. Figure 19.1a shows the method of gravity control, the controlling torque T_1 due to the displaced mass m being in opposition to the deflecting torque T_2. This method has the serious disadvantage that the position of the instrument is critical since any movement will take the needle from the scale zero when T_1 is zero. Also the instrument cannot be used in a horizontal position. Figure 19.1b shows spiral spring control. It has neither of the above disadvantages, and it has the important advantage that the controlling torque T_2 is proportional to the deflection θ. Figure 19.1c shows the method of air-damping by means of a piston moving in a curved damping chamber. Another form consists of vanes moving in a flat box. Figure 19.1d shows electromagnetic damping by eddy currents induced in an aluminium disc by the field of a small permanent magnet M. If the moving element is a simple coil in a strong magnetic field, eddy-current damping can be provided by winding the coil on a copper former.

Types of Indicating Instrument

Instruments used for measuring current, voltage and power are classified according to their principle of operation.
1. Electromagnetic   Ammeters and voltmeters
2. Electrodynamic   Ammeters, voltmeters and wattmeters
3. Electrostatic   Voltmeters
4. Thermal   Ammeters
5. Induction   Ammeters, voltmeters and wattmeters
6. Rectifier   Milliammeters and millivoltmeters

The Moving-iron instrument

Figure 19.2 shows an electromagnetic instrument whose operation depends on the repulsion of two adjacent pieces of iron magnetised by the same field. The irons A and B are tongue-shaped, A being fixed and B attached to the moving system. The poles of the strips are along the edges, as shown in the sectional view, and consequently repulsion takes place. The scale of the instrument is not uniformly divided, but by shaping the irons correctly, a very nearly uniform subdivision of the most useful part of the scale can be obtained, say from 20 per cent of full scale upwards. Since the polarity of the

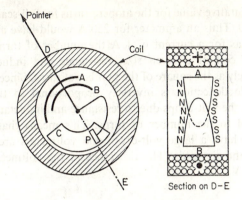

Figure 19.2   Moving-iron instrument.

strips reverses if the coil current is reversed, the instrument can be used for both direct and alternating current.

The sources of error are as follows:

(i) Hysteresis error which causes the readings with descending values of current or voltage to be higher than with ascending values. It can be reduced to a very small value by the use of a low-loss alloy such as mumetal (see chapter 8). As the relative permeability is of the order of 30 000 at very small magnetising forces the use of this material gives a high magnetisation of the irons and therefore a high torque.

(ii) Stray magnetic fields. The effect of a stray field is small when the magnetisation of the irons is strong, another reason for using mumetal. For accurate work the working part should be screened by an iron case or thin iron shield.

(iii) Frequency errors. These apply to a.c. working only. Since the impedance of a coil increases with the frequency, the current through a voltmeter coil will decrease with increased frequency, even if the p.d. is constant. A large non-inductive series swamping resistance is sufficient if the changes in frequency are likely to be small. The error can be reduced by connecting a capacitor in parallel with the series resistance since by a correct choice of C the whole circuit can be made independent of frequency.

A representative value for the ampere-turns for full-scale deflection is about 250. Thus an ammeter for 250 A would have a coil of one turn, whereas for a current of 1 A the number of turns would be 250. For a coil of a given size, resistance and inductance vary approximately as the square of the number of turns. Since the current for full-scale deflection is inversely proportional to the turns, it follows that the volt-drops due to resistance and inductance are each proportional to the number of turns. This means that low-range instruments have a large volt-drop, and voltmeters are preferably for 100 V and upwards. The resistance of the voltmeter is of the order 20 Ω/V.

The permanent-magnet moving-coil instrument

Figure 19.3 shows a rectangular coil in a radial magnetic field. The deflecting torque is independent of position and is proportional to I. With coil spring control the restoring torque is proportional to the deflection θ and since the coil comes to rest at such a value of θ that these two torques are equal it follows that the deflection is proportional to the coil current. This is exceedingly important since it means that the scale is uniformly divided.

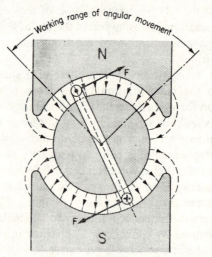

Figure 19.3 Coil in a uniform radial magnetic field.

The essentials of the instrument are (a) a rectangular moving coil, controlled by a spring and (b) a permanent magnet providing a uniform radial field in the air gaps within which the coil moves. The essentials of the instrument are therefore as in figure 19.4. Since the magnetic flux is unidirectional a reversal of current will reverse the deflection showing that the instrument cannot be used with alternating current in the coil.

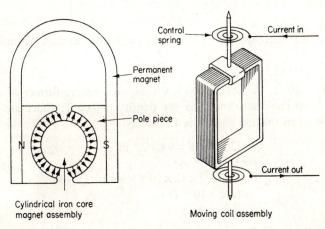

Figure 19.4 Essential parts of a permanent-magnet moving-coil instrument.

The torque is calculated as follows:

Let B = flux density in the air gaps over the range of movement
N = no. of turns
l = coil length and d coil width (figure 19.5)
I = coil current
Force acting on one side
$F = BIl$ N
∴ Deflecting torque
$T = Fd = BIldN$ N m
Let k = restoring torque of spring per radian deflection, then, when the system has come to rest
$k\theta = BIld$ N
$\theta = (BldN/k) \times I$

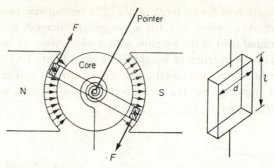

Figure 19.5   Arrangement of permanent magnet moving-coil instrument.

Example 19.1.   A moving coil is wound on a square former of side 3 cm, and has 100 turns. The gap density is 0.06 T. Calculate the torque when the coil current is 12 mA.

$$ld = (3 \times 10^{-2})^2 = 9 \times 10^{-4} \text{ m}^2$$
$$I = 12 \times 10^{-3} \text{ A}$$
$$\therefore \quad T = 6 \times 10^{-2} \times 12 \times 10^{-3} \times 9 \times 10^{-4} \times 10^2$$
$$= 6.48 \times 10^{-5} \text{ N m}$$

Extension of the Instrument Range

The moving-iron instrument can be used as either voltmeter or ammeter by a suitable design of coil. With the moving-coil instrument, the instrument itself is not changed when the range is changed. Usually it is one of coil resistance 5 Ω and giving a full-scale deflection when the coil p.d. is 75 mV. Thus the coil current for full deflection is

$$I_c = (75 \times 10^{-3})/5 = 15 \times 10^{-3} \text{ A or 15 mA}$$

Thus the instrument alone can be used as a voltmeter up to 75 mV, or an ammeter up to 15 mA.

For use as a voltmeter to measure up to any value V, the total resistance must be such that the current I_c flows, and therefore there must be a series resistor R to ensure this, (figure 19.6a). Calling the total resistance R_t we have

$$R_t = V/I_c$$
$$\therefore \quad R = (V/I_c) - R_c$$

For use as an ammeter to measure any value up to I, the instrument must be shunted by a resistor R_{sh}, figure 19.6b such that R_{sh} carries a current of $(I - I_c)$. Calling the instrument resistance R_c

$$\frac{R_{sh}}{R_c} = \frac{I_c}{I_{sh}} = \frac{I_c}{I - I_c}$$

Thus, by providing the same instrument with a number of series resistors and a number of shunts of different values, a very wide range of voltage and current can be covered.

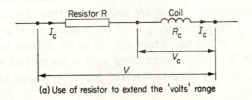

(a) Use of resistor to extend the 'volts' range

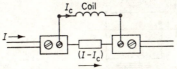

(b) Use of shunt to extend the 'amps' range

Fgure 19.6

Example 19.2. A moving-coil instrument given a full-scale deflection with a p.d. of 75 mV, the current then being 25 mA. Calculate (a) the series resistor which will give a range of 0–100 V, (b) a shunt which will give a range of 0–50 A.

(a) $$R = (V/I_c) - R_c$$

and $$R_c = \frac{75 \times 10^{-3}}{25 \times 10^{-3}} = 3 \ \Omega$$

$$\therefore \quad R = \frac{100}{25 \times 10^{-3}} - 3 = 3997 \ \Omega$$

(b) $$R_{sh} = \frac{R_c I_c}{I - I_c}$$

$$= \frac{3 \times 25 \times 10^{-3}}{50 - 25 \times 10^{-3}} = 0.001501 \ \Omega$$

The range of a moving-iron voltmeter can be increased by the use of a series resistor, but it is not usual to increase the range of a moving-iron ammeter by the use of shunts. This is because of the relatively large power loss in this type of ammeter. For use on a.c. circuits the range of a voltmeter can be increased by means of a voltage transformer and the range of an ammeter by means of a current transformer, as shown in figure 19.7. Because of the volt drops in the winding, the voltage ratios of the voltage transformer and the current ratios of the current transformer are not constant but are dependent on the load on the secondary side. Thus there are ratio errors which have been allowed for if the instruments are used for fine measurement. There are also phase errors due to the slight phase shift of the secondary p.d. of the voltage transformer and of the phase shift of the secondary current of the current transformer. These are of little importance when measuring current or voltage, but they can be very important when measuring power, as explained later.

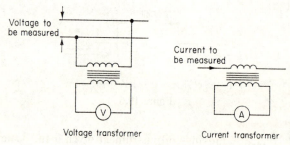

Figure 19.7   Instrument transformer connections.

Effect of instrument resistance

Figure 19.8a shows a simple circuit including an ammeter, voltmeter and resistor. It is assumed that the ammeter resistance is zero and the voltmeter resistance infinite (ideal but not attainable). Then the voltmeter reading $V = IR$ and $R = V/I$, so that resistance could be measured by this means. It is, in fact, the very convenient 'ammeter and voltmeter' method.

Figure 19.8b shows the effect of a finite voltmeter resistance. There is a finite current I_v, and the current through R, $(I - I_v)$ is less than the measured current. The volt drop is therefore $(I - I_v)R$ instead of IR, and if R is determined in this way, the calculated value will be

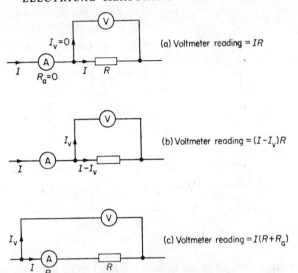

Figure 19.8 Effect of instrument connections.

$(I-I_v)/I$ of the correct value. The connection scheme is therefore suited to the measurement of low resistances capable of carrying large currents. (I_v) will then be very small and the error small.

Figure 19.8c shows one voltmeter lead taken to the far side of the ammeter. The voltmeter reading is now $I(R+R_a)$ instead of IR, and the calculated value of R will be too high in the ratio $(R+R_a)/R$. This connection is therefore suited to high values of R, the ammeter resistance R_a then being relatively very small.

Knowing the instrument data the necessary corrections can be calculated very easily, but whether they can be calculated or not, the best practice is to choose the connection scheme which gives the smallest errors.

The Dynamometer Instrument

This is an electro-dynamic instrument, the magnetic field being provided, not by a magnet, but by a pair of field coils. The construction is shown, in figure 19.9. Control is by a pair of helical springs, not shown, and air damping is usually employed, the device consisting of light vanes moving in a flat box. Two field coils are

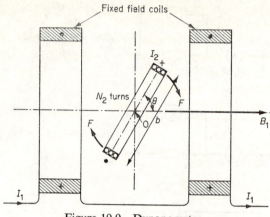

Figure 19.9 Dynamometer.

used because this gives an almost uniform field over the range of movement of the moving coil.

By having high resistance fine-wire coils and connecting fixed and moving coils in series, the instrument can be used as a voltmeter.

If the direction of the current through the coils is reversed, the direction of the magnetic field is also reversed, and the torque is therefore in the same direction as before. Thus the instrument is suitable for both d.c. and a.c. circuits. If somewhat thicker wire is used the instrument can be used as an ammeter, but there is a limit to the current range because of the difficulty of leading heavy currents to and from a moving coil. The difficulty can be partly overcome by connecting the moving coil to the terminals of a shunt. Connections for use as a voltmeter and an ammeter are given in figure 19.10.

The strength of the magnetic field is proportional to the current I_f in the fixed coils because no iron is used. Hence if I_m is the current in the moving coil the torque on this coil is proportional to the product $I_f I_m$. If these two are equal as in the voltmeter and if $I_m \propto I_f$ as in a shunted ammeter the torque is proportional to the square of the current (showing that when used as an ammeter or voltmeter the instrument has a scale unevenly divided).

The most important use of the dynamometer instrument is as a wattmeter the connections, without instrument transformers, being those of figure 19.10c. Different voltage ranges can be accommodated by means of an external variable resistor, as in the case of a moving-coil or moving-iron voltmeter. The field strength is proportional to I, the line current, and the moving-coil current is proportional to the voltage V. The torque is thus proportional to the product VI, which is the power in the d.c. case, and the wattmeter therefore has a uniformly divided scale. Since there is no iron the field strength is low and therefore magnetic shielding should be provided by enclosing the movement in an iron case.

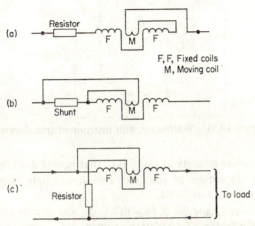

Figure 19.10 Scheme of connection of a dynamometer instrument when used as (a) a voltmeter, (b) an ammeter, (c) a wattmeter.

If used on a.c. circuits the instrument still measures the power for let

$$v = V_m \sin \omega t$$
and $i = I_m \sin (\omega t - \phi)$ where $\cos \phi$ is the p.f., then

average of $vi = V_m I_m \times$ average of $\sin \omega t \sin (\omega t - \phi)$
$$= \tfrac{1}{2} V_m I_m \cos \phi$$
$$= VI \cos \phi = P$$

Thus, in the alternating case also the deflection is proportional to the power.

On a.c. circuits it is usual to increase the range by means of voltage and current transformer as shown in figure 19.11.

The above simple theory ignores the fact that the voltage coil circuit possesses inductance, although small, and therefore the current in it lags the voltage by a small angle, θ say. If instrument transformers are used these, also, will cause phase differences, and if we denote by θ the total of all these effects it can be shown that the wattmeter reading has to be multiplied by a correction factor of $\cos \phi / [\cos \theta \cos (\phi - \theta)]$.

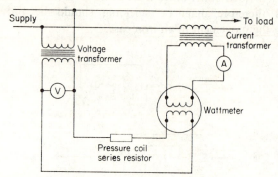

Figure 19.11   Wattmeter with instrument transformers.

It is very close to unity at unity p.f., that is, $\cos \phi = 1$, but there is appreciable departure at low p.f. As an example, suppose that $\theta = 2°$ so that $\cos \theta = 0.9994$.

At unity power factor $\phi = 0$, $(\phi - \theta) = -2°$, $\cos (\phi - \theta) = 0.9994$

$\therefore$ correcting factor $= 1/(0.9994)^2 = 1.0012$

the error therefore being 0.12 per cent

Let the p.f. be 0.5, then $\phi = 60°$ and $(\phi - \theta) = 58°$, $\cos (\phi - \theta) = 0.530$

$\therefore$ correcting factor $= 0.5/0.9994 \times 0.530 = 0.943$

the error now being 5.7 per cent. We see that for accurate power measurement, particularly at low power-factors, the wattmeter and instrument transformers must be of accurate laboratory type in order to ensure that the angle θ is very small.

Thermo-junction instruments

A thermo-junction is a junction of two dissimilar metals. If such a junction is heated an e.m.f. will be produced and therefore current will flow if the junction is part of a closed circuit. Figure 19.12 shows the principle, from which it will be seen that it is a permanent magnet, moving-coil instrument with the thermo-junction J as the source of moving-coil current. The junction is heated by means of a coil F which carries the current to be measured. The instrument in the figure is of the non-contact or convection type in which both junction and heater are in air. In the non-contact vacuum type the junction is attached to the heater by a vitreous bead but there is no electrical contact. This type is suitable for currents up to 150 mA and the convection type for currents greater than this. There is a third type, the contact type, in which there is electrical connection between heater and junction, the advantage being that there is a more rapid response to change in heater current than with the others.

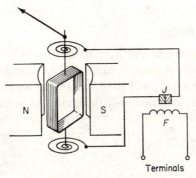

Figure 19.12   Thermo-junction instrument.

The instrument is suitable for small currents, both direct and alternating. The reading is entirely independent of waveform and of frequency, for which reason its most important field of application is the measurement of high-frequency currents. The thermo-couples are iron–eureka, copper–eureka and, for special purposes, platinum–rhodium. For a temperature of 200° C above the rest of the circuit, the thermo-electric e.m.f.s are, respectively, 11, 9.3 and 1.43 mV. The heaters are almost pure resistances, even at high frequencies, the material usually being a high resistivity alloy, although carbon can be used for very small currents.

Rectifier Instruments

Obviously for alternating current only. The circuit is that of figure 18.25c, a moving coil, permanent-magnet instrument taking the place of the load resistor. The bridge circuit, which gives full-wave rectification, is necessary because zero torque for the whole period of a half-wave is inadmissible. The reading of the instrument is not independent of waveform so that the scale can only be graduated in r.m.s. values if the current to be measured is sinusoidal. Actually, the instrument indicates the average current and therefore, if this is not sinusoidal, the waveform must be known if the r.m.s. value is required. The power consumption is very small, and as the whole circuit is series connected (being an ammeter) there is no high peak reverse voltage to withstand. Hence copper-oxide and selenium rectifiers are used, the former being preferable because of its slightly more linear characteristic.

The Wheatstone Bridge

Imagine two resistors R_1 and R_2 in parallel, as in figure 19.13. There will be a voltage drop along R_1 and the same voltage drop along R_2. Some point C along R_1 will have a potential intermediate between those of A and B, and obviously there must be some point D along

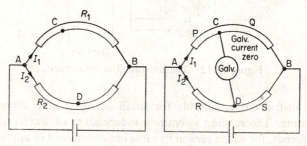

Figure 19.13   The Wheatstone bridge.

R_2 which has the same potential as C. If these two points are joined by a wire no current will flow along the wire and the currents along R_1 and R_2 will be unchanged. If C and D are joined to the terminals of a galvanometer there will be no deflection. Under these conditions

Drop along AC = Drop along AD
Drop along CB = Drop along DB

Call the resistances of the two portions of R_1, P and Q, and of R_2, R and S. Then

$$PI_1 = RI_2 \text{ and } QI_1 = SI_2$$
$$\therefore P/Q = R/S \text{ or } R = (P/Q) \times S$$

Thus, if P, S and Q are known, R can be calculated.

To use this method of determining an unknown resistance R, the ratio P/Q must be variable, or S must be variable, or both. In the metre bridge, figure 19.14, it is the ratio P/Q which is variable, and to achieve this the branch ACB (the lettering corresponds to figure 19.13) is in the form of a length of resistance wire, such as german silver, stretched along a metre scale. There is a sliding contact C, so that the point C can be anywhere along AB, the ratio P/Q thus being variable over wide limits. Two copper bars are used as the points A and B, and there is a third bar placed between them so as to provide two gaps in which a standard resistance S, and the

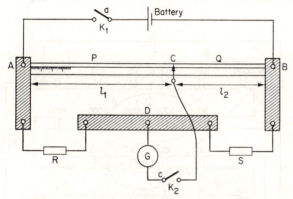

Figure 19.14   Connections of the metre bridge.

unknown resistance R can be placed. It is usual to have a key K_1 in the battery circuit, a convenient battery being a Leclanché cell (not an accumulator) while the sliding contact C is itself a key, represented diagrammatically by K_2. The method is to adjust C until a balance is obtained, as indicated by a zero galvanometer deflection. If l_1 and l_2 are the lengths of wire on either side of C then

$$R = (l_1/l_2)S$$

If S can be varied, or there is a choice of fixed standard resistor, it is a good plan to choose S such that C comes about the middle of the wire. This gives the greatest freedom from error. Suppose, for example, there is an error in reading the position of C of 0.5 mm, or 0.05 cm. Then, if the balance point should be in the middle and l_1/l_2 therefore unity, the observed value will be 49.95/50.05. an error of one part in 500, or 0.2 per cent. If, however, the balance point is at, say, $l_1 = 90$, $l_2 = 10$, then the same observational error of 0.05 cm gives an observed ratio of 89.95/10.05 or 8.95 instead of 9.0. The error is now 0.05 in 10, or one part in 200, or 0.5 per cent.

The Post Office Box

This is an accurate form of Wheatstone bridge in which the slide-wire is replaced by variable ratio arms P and Q of the plug-box type, and the standard S can be varied in units from one to several thousand ohms. How this is done will be clear from figure 19.15, in

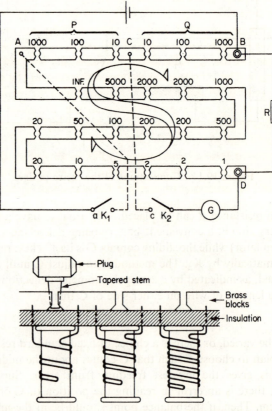

Figure 19.15 Connections and construction of coils in the Post Office box.

which the lettering is the same as in the previous figures, so that the circuit can be compared with figures 19.13 and 19.14. In the figure, the connections shown dotted are part of the instrument itself, while the others have to be made by the experimenter. There is one difference from the previous figures, namely, that R and S are interchanged so that at balance we now have

$$P/Q = S/R, \; R = (Q/P)S$$

The method of using the bridge is as follows: P and Q are made equal to one another, say 100 Ω each, and the plug marked infinity withdrawn; the plug has no resistance across its gap so that its removal open-circuits S. The galvanometer is shunted by a low resistance if necessary, and the key then depressed, first K_1, then K_2, the latter by a momentary touch only. The direction of the galvanometer deflection is noted, and whenever a deflection in the same direction is observed, it means that the resistance in S is too large. A deflection in the opposite direction means that the resistance in S is too small. The infinity plug is now replaced and the plugs in S adjusted until a certain resistance gives a throw in the 'too small' direction and 1 Ω more gives a throw in the 'too large' direction. This means that R lies between these two values, for example, they might be 54 and 55.

The value of P is now increased to 1000, leaving Q at 100, and the above process repeated. When the nearest balance is obtained, this time it will be necessary to increase the sensitivity of the galvanometer by increasing the shunt resistance. Since P/Q is now equal to 10, the value of S to give a balance will be ten times what it was before. Suppose that 546 is now too small and 547 too large, then the value of R lies between 54.6 and 54.7.

The ratio P/Q is now made 1000/10 that is 100/1 so that for a balance S must be 100 times R. We can again obtain two different values of S differing by unity. Let them be 5463 and 5464 then the value of R lies between 54.63 and 54.65 ohms. The exact value is now determined by using the maximum sensitivity of the galvanometer by removing the shunt and observing the magnitudes of the steady deflections in the two directions; first with S equal to 5463 and then with S equal to 5464. Suppose, for example, that these two deflections are 18 and 29 divisions respectively, then the difference of 1 Ω corresponds to a total deflection of $18 + 29$, so that the final value of R is

$$R = \frac{5463 + 18/(18+29)}{100} = 54.634 \text{ ohms}$$

The metre bridge is a variable ratio bridge, and the post office box a fixed ratio and variable standard bridge. This is because the manipulations are made with a fixed ratio P/Q.

Galvanometers and Shunts

A galvanometer is an instrument used for measuring, or detecting, exceedingly small currents. They can be used for deflectional methods, but often they are used to find a balance, as in the Wheatstone bridge circuit, by indicating zero current. The moving-coil galvanometer is a modification of the permanent-magnet, moving-coil indicating instrument designed to give the maximum sensitivity as expressed by the deflection per microampere. The constructional features are as shown in figure 19.16.

(1) The suspension in bearings and control by spiral springs are replaced by a thin length of phosphor bronze strip. This provides the controlling torque and one connecting lead. The other lead is a ligament at the bottom of the coil. The sensitivity depends on the length of the suspension and also on its cross-section, and very high sensitivities can be obtained by using a long fine strip.

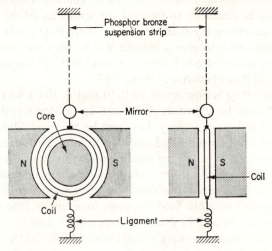

Figure 19.16   Types of moving-coil galvanometer.

(2) The coil is similar to that of the indicating instrument except that both circular and rectangular forms are used. With coils of normal proportions an iron core to concentrate the magnetic field can be used, but this is not possible if the coil is long and narrow. Such coils have a small moment of inertia and therefore their time of swing is short, often an advantage, particularly with 'null' methods. If damping is required the coil can be wound on a metal former, as explained previously. Alternatively, when a deflection has been read, the coil can be short-circuited by means of a key connected directly across the terminals. The movement of the coil then induces current in the coil itself, thus providing eddy-current damping.

(3) Deflection. Instead of a pointer moving over a scale, a beam of light is reflected from a small mirror attached to the movement so as to form a spot of light on a translucent scale. The distance from mirror to scale is usually one metre, but since the angular deflection of the beam is twice the angular deflection of the mirror this gives, in effect, a pointer two metres long.

The sensitivity can be varied by shunting the moving coil, very like the shunting of an ammeter. Suppose that the galvanometer is to carry only $1/m$ of the total current. Then if we denote the total current by m units, the galvanometer must carry 1 unit. The shunt must therefore carry $(m-1)$ units. Since shunt and galvanometer are in parallel, then, denoting galvanometer and shunt resistances by G and S respectively.

$$\frac{S}{G} = \frac{\text{Galv. current}}{\text{Shunt current}} = \frac{1}{m-1}$$
$$\therefore \quad S = G/(m-1)$$

Thus if the galvanometer is to carry 1/10th of the current $S = G/9$. If it is to carry 1/100th, $S = G/99$, and so on.

EXERCISES

Chapter 1

Take 1 lb = 454 g, 1 in = 2.54 cm

1. If a mass of 10 lb is lifted a vertical distance of 75 ft calculate the work done against gravity. *Ans* 203.5 J

2. A road has a slope of 8°. A lorry of gross mass $3\frac{1}{2}$ tons is driven for one mile up the road. Calculate the work done against gravity.
   Ans 7.88×10^6 J

3. If a force of 25 lb wt acts on a body of 200 lb what will be the acceleration produced in SI units. *Ans* 1.228 m/s²

4. What applied force is required to impart to a mass of 1000 lb an acceleration of 5 ft/s². *Ans* 691 N

5. Two forces of 30 N and 50 N respectively act at a point (a) in the same direction (b) in opposite directions, (c) at right-angles to one another. Calculate the resultant force in each case. *Ans* 80, 20 and 58.3 N

6. A lever AB, 100 cm long is balanced at its mid-point. A mass of 20 g is attached at A. (a) What is the moment of the force produced by it? (b) What mass must be attached 15 cm from the end B to balance the lever? *Ans* (a) 0.098 N m, (b) 28.5 g

7. A motor driven hoist has a drum diameter of 2 m and a drum speed of 75 rev/min. If the weight lifted is 250 kg, calculate the power required. What is the work done when the weight has been lifted a height of 25 m? *Ans* 19 250 W; 61 310 J

8. The weight of a grandfather clock is 15 lb. In one week it descends 4 ft 6 in. How many watts are required to drive the clock?
   Ans 1.52×10^{-4}

9. An electric motor has an intake of 17 600 W and the losses in it are 2000 W. If the motor speed is 1000 rev/min what torque does it develop? *Ans* 149.1 N m

287

288 BASIC ELECTROTECHNOLOGY

10. A generator supplies 50 kW at a speed of 720 rev/min. Its efficiency at this loading is 85 per cent. Calculate the torque which is applied to the shaft. *Ans* 779 N m

11. An electrically driven hoist raises a mass of 1 cwt through a height of 25 ft in 45 s. The overall efficiency of motor and hoisting gear is 54 per cent. What are the output and intake of the motor? *Ans* 14.55hp; 27hp

12. An electric motor drives an electric generator. The efficiency of the motor is 80 per cent and that of the generator 75 per cent. If the output of the generator is 6 kW, what is the intake of the motor? *Ans* 10 kW

13. In a certain piece of equipment the force on an electron, of mass 9.1×10^{-28} g is 1.6×10^{-15} N. If it starts from rest calculate its acceleration, the time taken to travel 1 cm and its kinetic energy at that position. *Ans* 1.8×10^{15} m/s, 6.0×10^6 m/s, 3.3×10^{-9} s, 16×10^{-18} J

Chapter 3

1. A metal sphere of 5 cm radius has a charge of 10×10^{-9} C. What is the potential due to this charge at a point very close to the surface? Plot the graph of potential against distance from the centre. *Ans* 1800 V

2. Charges of $+20 \times 10^{-9}$ and -20×10^{-9} C are placed 10 cm apart. Calculate the potentials (a) at a point in the line of the charges 4 cm from the negative charge, (b) at a point in the line of the charges 4 cm outside the positive charge. *Ans* (a) -1500 V, (b) $+3220$ V

3. If two particles each with a charge of 3.2×10^{-19} C repel one another with a force of 9.18×10^{-2} N. What is their distance apart? *Ans* 10^{-9} m

4. If a 200V battery is connected to two large parallel plates 1 cm apart, what is the electric field strength between them? *Ans* 2×10^4 V/m

5. The radius of the orbit of the single electron of the hydrogen atom, is 5.28×10^{-19} cm. Since the charge in the electron and therefore in the nucleus is 1.6×10^{-19} C what is the field strength of the field in which the electron rotates? *Ans* 5.18×10^7 V/m

6. Three positive charges of 1, 2 and 3×10^{-9} C are placed at the corners of an equilateral triangle of side 1 m. Calculate the potential and the electric field strength at the centre of the triangle. *Ans* 93.5 V, 46.8 V/m

7. If the maximum electric field strength at the surface of a charged conductor is 3×10^6 V/m, above which value it will discharge by imparting its charge to the air molecules, what is the maximum potential to which a metal sphere of radius 3 m can be raised? *Ans* 9×10^6 V

Chapters 4, 5, 6 and 7

1. There are three resistances in series, their values being 1, 2 and 3 Ω respectively. If they carry a current of 5 A what will be the potential differences (a) across each resistor, (b) across all three?
Ans (a) 5, 10 and 15 V, (b) 30 V

2. A cell of e.m.f. 2.2 V and resistance 0.5 Ω is connected to a coil of resistance 1.5 Ω. Calculate the current and the p.d. between the cell terminals. *Ans* 1.1 A, 1.65 V

3. The terminal p.d. of a battery of cells is 18 V when supplying a current of 4A and 17 V when supplying 6 A. Calculate the e.m.f. and resistance of the battery. *Ans* 20 V, 0.5 Ω

4. A cell of e.m.f. 1.5 V and resistance 1.0 Ω is connected to two coils in series, one of 0.5 Ω, the other of 0.75 Ω. Calculate the volt-drops across each. *Ans* 0.33 V, 0.5 V

5. Two wires of 2 and 3 Ω respectively are connected in parallel. What is their joint resistance, and what will be the branch current if they carry a total of 10 A. *Ans* 1.2 Ω, 6 and 4 A respectively

6. Four wires each of 1 Ω resistance are arranged to form a square. Calculate the resistance (a) between adjacent corners, (b) between opposite corners. *Ans* (a) 0.75 Ω, (b) 1.0 Ω

7. A resistance of 3 Ω is in parallel with one of 4 Ω, and the two are in series with a resistance of 5 Ω. This circuit is supplied from a battery of e.m.f. 10 V and resistance 0.5 Ω. Calculate the current, the battery p.d. and the drop in volts across each resistance.
Ans 1.385 A, 9.3 V, 2.37, 2.37 and 6.93 V

8. The field winding of a certain series motor has a resistance of 0.2 Ω the current through it being 20 A. For the purpose of reducing the current through this winding a 'diverter' resistance of 0.5 Ω is connected in parallel with it. Calculate the combined resistance of coil and diverter and the current through each.
Ans 0.143 Ω; 28.6 A and 11.4 A

9. A wire is formed into a closed circle of 30 cm diameter, and the wire is such that each cm has a resistance of 0.1 Ω. At two points one-third of a circumference apart thick wires of negligible resistance are connected to a battery of e.m.f. 15 V and resistance 5 Ω. Calculate the battery current and terminal p.d. and the currents in the two parts of the wire. *Ans* 2.12 A, 4.4 A, 0.706 and 1.412 A

10. Three resistors of 4, 6 and 12 Ω are connected in parallel and supplied from a cell of e.m.f. 1.5 V and resistance 1 Ω. How many coulombs are delivered by the cell in one minute? *Ans* 30 C

11. A cell of constant e.m.f. is connected to a resistor of 2 Ω and a current of 1.1 A flows. If the 2 Ω resistor is removed and replaced by one of 10 Ω, the current is 0.3 A. Calculate the internal resistance of the cell. *Ans* 1 Ω

12. A battery of e.m.f. 20 V and resistance of 0.5 Ω is connected to a circuit consisting of a resistance of 1 Ω in series with three resistances of 2, 3 and 4 Ω in parallel. Calculate the total current, the current in the three parallel branches, and the terminal p.d. of the battery.

Ans 8.20 A; 3.81 A; 2.54 A and 1.90 A; 15.88 V

13. Three resistors are joined to form an equilateral triangle ABC. AB = 1 Ω, BC = 2 Ω, CA = 3 Ω. A cell of e.m.f. 2 V and resistance 1 Ω is connected to points A and C. Calculate all the currents and the cell p.d.   *Ans* 0.4 A and 0.8 A; 1.2 V

14. There are two wires A and B of the same metal. A is 20 times as long and one-third of the cross-section of B. If the resistance of A is 1 Ω, what is the resistance of B?   *Ans* 1/60 Ω

15. There are two wires A and B. A is ten times as long as B and is half the cross-section of B. The resistivity of A is three times that of B. Calculate the ratio of the resistances.   *Ans* 60:1

16. The resistivity if manganin is 440 nΩ m. What length of wire of diameter 0.1 mm will be required to make a coil of 1000 Ω resistance?
                                                            Ans 17.9 m

17. Find the resistance of a wire 200 m long and of circular cross-section of diameter 0.5 mm. The resistivity of the material is 1.6 MΩ cm.
                                                            Ans 16.33 Ω

18. A wire of length 6 m and diameter 0.5 mm has a resistance of 1.5 Ω. What is its resistivity?   *Ans* 4.91 MΩ cm

19. A battery supplies 250 identical resistors in parallel, each of 300 Ω. The p.d. is then 120 V. It rises to 122 V when 100 of the resistors are switched off. Calculate the battery resistance.   *Ans* 0.051 Ω

20. A wire carries a current of 10 A. How many electrons pass any cross-section in 10 s?   *Ans* 0.62×10^{21}

21. A mining haulage road signalling system consists of two bare galvanised wires 200 yd long and of cross-section 0.02 in². The battery used has an e.m.f. of 24 V and a resistance of 5 Ω; the bell has a resistance of 10 Ω. Calculate the current if the wires are brought into contact with one another half-way along the road. Resistivity of the wire 4.7 MΩ in.   *Ans* 1.44 A

22. A haulage road cable 1½ miles long has cores of 0.1 in² cross-section. How will the resistance of each core change if its temperature increases from 15° C unloaded to 50° C when carrying current. Take α = 0.0041.
                                                            Ans 0.094 Ω

23. The field current of a 500 V motor was 2.38 A when cold at 20° C, and 2.17 A after the motor has been running for some hours. Assuming α = 0.00393 for the initial temperature of 20° C calculate the average temperature rise of the winding.   *Ans* 24.3° C

24. The resistance of a certain conductor is 5.0 Ω at 20° C and 7.2 Ω at 100° C. What is its temperature coefficient?   *Ans* 0.0055 at 20° C

25. A certain substance has a negative temperature coefficient of 0.01. If a rod of this substance has a resistance of 1 Ω at 0° C, what resistance of copper wire of temperature coefficient 0.0039 at 0° C must be connected in series with it to make the resistance of the two independent of temperature? *Ans* 3.56 Ω

26. A cell of e.m.f. 1.1 V and resistance 1Ω is joined to a galvanometer of resistance 50 Ω. This is shunted by a wire of diameter 0.5 mm and length 20 cm, resistivity 4.4×10^{-5} Ω cm. Calculate the galvanometer current. *Ans* 0.00677 A

27. A certain filament lamp has a resistance of 21 Ω at 0° C and 31.2 Ω at 100° C. When operating at its normal voltage its resistance is 250 Ω. Assuming a linear relationship between resistance and temperature, calculate the temperature during normal operation. *Ans* 2.245° C

28. A 600W electric kettle is designed for a 220V supply. What current does it take and what is its resistance? *Ans* 2.73 A, 80.6 Ω

29. A current of 5 micro-amps passes through a coil of 0.2 MΩ. What is the power? *Ans* 5×10^{-6} W

30. A certain grade of resistance wire has a resistance of 1360 Ω per 1000 yd. What length of this wire would be required for a heater taking 400 W at 220 V? *Ans* 88.95 yd

31. There are three coils in parallel, the values being 2, 2.5 and 4 Ω respectively. If a total current of 10 A is divided between them, calculate the power in each. *Ans* 37.8, 30.2 and 18.9 W

32. There are two conductors A and B. A is ten times as long as B, but is of one-third the cross-section of B. Calculate the ratio of their resistivities. *Ans* 1:1

33. There are two conductors A and B. A is twice as long as B but is one-half the cross-section of B. If the two are connected in parallel to a given supply, the power in B is twice that of A. Calculate the ratio of their resistivities. *Ans* 1:2

34. A p.d. of 10 V is applied to two coils in parallel. If the total current is 4 A and the power in one coil is 15 W, calculate the resistance of each coil. *Ans* 6.67 Ω and 4 Ω

35. A generator delivers 250 A at 600 V. What is its output in kW? If its efficiency at this load is 92 per cent, calculate the hp output of the engine driving it. 1 kW = 1.34 hp. *Ans* 219 hp

36. A small 220 V motor has an efficiency of 40 per cent when developing $\frac{1}{8}$ hp. What current does it take? How much will it cost to run the motor with electricity at 1$\frac{1}{2}$p per unit? *Ans* 1.06 A; 0.35p

37. The motor of a large reversing mill develops 20 000 hp, its efficiency then being 96 per cent. What current does it take if supplied at 600 V? *Ans* 25 900 A

38. In a certain house there are, on the average throughout the year, six 60 W lamps in use for five hours per day. If one unit of electricity for

lighting purposes costs 3p, what is the average quarterly bill for lighting? *Ans* £2.05

39. A pump driven by a 500 V d.c. motor lifts 100 000 gallons of water per hour to a height of 150 ft. The efficiency of the pump is 80 per cent and of the motor 90 per cent. Calculate the current taken. If the pump runs for at 8 hours per day and one unit costs $\frac{1}{2}$p, calculate the units consumed and the cost per day. *Ans* 631, £3.15$\frac{1}{2}$p

40. Find the cost of raising 3 pints of water from 15° C to boiling point in an electric kettle whose efficiency is 85 per cent. If the heating element takes 500 W, how long will it take? The cost of one unit of electricity is 1p. *Ans* 0.2p, 23$\frac{1}{2}$ min

41. A battery of e.m.f. 20 V and resistance 1 Ω has a resistance of 5 Ω connected to its terminals. How many kcal will be generated in the wire in 10 minutes? *Ans* 7.9

42. A battery of e.m.f. 10 V and resistance 2 Ω is in series with a 2 Ω resistor and current flows for half an hour. Calculate the heat generated in the resistor and in the battery. *Ans* 4.4 and 1.76 kcal

43. A coil of wire is immersed in 0.1 kg of water and a current of 1 A passed through it. It is found that the temperature of the water rises at the rate of 5° C per minute. What is the resistance of the coil? *Ans* 34.7 Ω

44. An electric heater supplied at 220 V takes 6000 W. Calculate (a) the heater resistance, (b) the electrical energy consumed per minute, (c) the mass of water which could be raised from 15° C to boiling point in 5 minutes, assuming negligible heat loss. *Ans* (a) 8.07 Ω, (b) 0.1 kWh, (c) 5.06 kg

45. A coil of resistance 10 Ω is immersed in two litres of water at 20° C. The coil is connected to a battery of e.m.f. 100 V and resistance 2.5 Ω. Assuming that 15 per cent of the heat produced is lost by radiation, how long will it be before the water is raised to 80° C? *Ans* 15.25 min

46. An electric cable one mile long has two cores each of 0.1 in² cross-section. It is supplied at 500 V. If the cable is suddenly short-circuited at the far end, calculate the initial rate of rise of temperature of the copper cores, given that the resistivity of copper at the initial temperature is 0.67 $\mu\Omega$ in the specific heat of copper 0.1 and the specific gravity of copper 8.8 *Ans* 23.2° C/min

47. A current of 2 A is passed through a coil of 0.706 Ω which is immersed in a lagged calorimeter. The mass of water is 42 g and the water equivalent of the calorimeter is 8 g. It is found that the temperature rises 6° C in 7 minutes. Calculate the value of J from this data. *Ans* 4.1×10^3. J/kcal

48. There are four cells each of e.m.f. 1.5 V and resistance 0.5 Ω. They are used to supply current to a 2.5 Ω resistor. What will this current be if they are connected (a) in series, (b) in parallel, (c) in series-parallel? *Ans* 1.333, 0.571 and 1.0 A

49. How many cells, each of e.m.f. 2 V and resistance of 0.5 Ω, when connected in series will be required to give a current of 2 A in a resistor of 3 Ω? *Ans* 6 cells

50. Eighteen cells, each of e.m.f. 1 V and resistance 2.2 Ω are to supply a circuit of 4 Ω resistance. What arrangement will give the maximum current, and what is this current? *Ans* 3 rows of 6 cells per row; 0.72 A

51. A battery consists of 24 cells each of e.m.f. 2 V and resistance 0.5 Ω. The resistance of the external circuit is 1.0 Ω. Plot a graph of current against cells in series per row and hence show, that since the number must be integral and divisible into 24, the possible numbers are 6 and 8.

Chapters 9 and 10

1. A straight wire carrying 100 A in a downward direction is arranged vertically in a uniform magnetic field of strength 10^{-3} T directed from south to north. Calculate the resultant field strength 5 cm from the centre of the wire (a) as the east side, (b) as the west side.
Ans (a) 6×10^{-4} T, (b) 1.4×10^{-3} T

2. There are two long parallel wires 10 cm apart. One carries a current of 50 A, the other of 25 A. Calculate the magnetic field strength at a point midway between them (a) with the currents in the same direction, (b) in opposite directions. *Ans* (a) 1×10^{-4}, (b) 3×10^{-4} T

3. A small motor has an air gap flux of strength 0.4 T. If an armature conductor is 1.5 cm from the centre, is 2 cm long, and carries a current of 1 A, calculate (a) the force acting on a conductor, (b) the moment of this force. *Ans* (a) 8×10^{-3} N, (b) 1.2×10^{-4} N m

4. A straight conductor 1 yd long and carrying 100 A is arranged at right-angles to a uniform magnetic field. If the force acting on the conductor is 5 lb wt what is the field strength? *Ans* 0.2213 T

5. A ring-shaped solenoid of mean circumference 40 cm and cross-section 5 cm² is wound with 200 turns. Calculate the flux produced by a current of 5 A. *Ans* 1.57×10^{-6} Wb

6. A non-magnetic ring of 10 cm mean diameter has a circular cross-section of 0.5 cm radius. It is uniformly wound with 500 turns of insulated wire. Calculate the magnetic flux produced by a current of 2 A. (Note that since there are no ends there are no end effects and the coil can therefore be regarded as a solenoid of length equal to the mean circumference.) *Ans* $\pi \times 10^{-7}$ Wb

7. A non-magnetic ring of mean diameter 1 ft is made from round material of diameter 1 in. If it has a coil of 2000 turns what current will produce a flux of 10^{-5} Wb? *Ans* 7.55 A

8. A round bar iron of diameter 2 cm and 100 cm long is placed inside a

solenoid 100 cm long, of 1000 turns, and carrying a current of 0.5 A. If the relative permeability of the iron is 750, calculate the flux produced.
Ans 1.48×10^{-4} Wb

9. An iron rod of circular cross-section 10 cm² is formed into a closed ring of mean circumference 100 cm. It is wound with 500 turns and a current of 3 A produces a flux of 2×10^{-3} Wb. What is the relative permeability of the iron?
Ans 1061

10. An iron ring has a mean diameter of 25 cm and a cross-section of 3 cm². An air gap of 0.4 mm is made by a radial cut across the gap in one place. The ring has a magnetising coil of 500 turns. Neglecting magnetic leakage and assuming a relative permeability of the iron of 2470, what current will be required to produce a flux of 2.1×10^{-4} Wb?
Ans 0.8 A

11. An iron ring of mean diameter 25 cm and cross-section 5 cm² has a single gap of 1 mm. If the magnetising coil has 4000 ampere-turns and the flux density produced in the gap is 1.1 T, calculate the relative permeability of the iron.
Ans 220

12. An iron ring of 100 cm mean circumference and 5 cm² cross-section has a magnetising coil of 500 turns. The iron is such that the following relationships exist:

Flux density T	1.02	1.2	1.37
Relative permeability	2000	1500	1000

If the ring is required to carry a flux of 6.5×10^{-4} Wb, what must be the current?
Ans 1.8 A

Chapter 11

1. A slow-speed d.c. generator has 16 poles. Its armature diameter is 3 m and the active length of each conductor is 0.51 m. If the flux density in the air gaps has an average value of 0.8 T and the speed is 100 rev/min, calculate the e.m.f. induced in each conductor.
Ans 6.4 V

2. A metal disc of radius 5 cm is rotated at 1200 rev/min in a magnetic field of 0.2 T, the direction of the field being perpendicular to the plane of the disc. Calculate the e.m.f. acting along a radius of the disc. (Note the speed at any point of the disc is proportional to its distance from the centre, hence the mean e.m.f. is one-half of the e.m.f. at the edge.)
Ans 0.0314 V

3. A conducting bar 5 ft long is resting on the floor of a guard's van of a train at a place where the vertical component of the earth's magnetic field is 2.2×10^{-5} T. If the train moves at 40 mile/h, what e.m.f. will be induced in the bar?
Ans 0.0006 V

4. A conductor of length 1.2 m moves with a velocity of 2 m/s at right-angles to a magnetic field of 2 T. Calculate the e.m.f. induced. If the ends of the wire are joined in such a way that no e.m.f. is induced in

the connecting wires, the circuit resistance being 2.4 Ω, what current will flow?   *Ans* 4.8 V, 2 A

5. A short piece of copper wire projects like a tooth from the rim of a wheel of 2 m diameter. The plane of the wheel is east-west at a place where the horizontal component of the earth's magnetic field is 1.8×10^{-5} T. If the wheel makes 30 rev/min, calculate the induced e.m.f. per cm length of the wire.   *Ans* $0.565 \times 10^-$ V

6. A certain electromagnet produces a flux of 0.01 Wb. The coil of the magnet has fifty turns. If the current is switched off and the flux becomes zero in 0.01 s, what will be the average e.m.f. induced in the coil?   *Ans* 50 V

7. A large electromagnet has a coil of 2000 turns. At a certain instant the flux is 2.5×10^{-3} Wb and 0.001 s later it is 2.25×10^{-3} Wb. Calculate the average e.m.f. induced in the winding during this interval.

                                                        Ans 500 V

8. The current in a coil of resistance 10 Ω and inductance 1 H increases uniformly at the rate of 10 000 A/s. Find the value of the necessary applied p.d. (a) when the current is 10 A, (b) when it is 50 A.
                                         Ans (a) 10 100 V; (b) 10 500 V

9. An iron ring of relative permeability 1333 is formed into a ring of mean circumference 100 cm and cross-section 10 cm². It has a magnetising coil of 600 turns. Over this is wound a secondary coil of 5000 turns. If the magnetising current of 2 A is switched off and the flux dies away in 0.01 s, what will be the average e.m.f. induced in the secondary winding?   *Ans* 1005 V

10. A 100 kW, 125 V shunt generator has 12 poles. Each pole has a magnetising coil of 650 turns and a magnetising current of 12 A produces a flux per pole of 8.5×10^{-2} Wb. Calculate the self-inductance of the field winding.   *Ans* 18.4 H

11. If the field winding of the generator of example 10 carries no current initially and is suddenly switched on to a 125 V supply, determine by graphical construction the time taken for the current to attain (a) $\frac{1}{10}$ (b) $\frac{1}{2}$ of its full value, given that the resistance is 10.4 Ω.
                                         Ans (a) 0.187 s, (b) 1.21 s

12. A RL circuit has $R = 1$ Ω and $L = 1$ H. The applied p.d. is 1 V. Plot the curve of current against time from 0 to 4 s, and show that after 2 s, the current will be 0.816 A.

13. A RL circuit has $R = 120$ Ω and $L = 2$ H. It is carrying a current of 5 A. If the circuit is suddenly closed on itself, plot the curve of decay of current and from it show that the time taken to fall to 0.25 A is 0.05 s.

14. There are two neighbouring coils A and B. When a current of 1 A flows in A the flux from A which links with B is 10^{-4} Wb. If B has 500 turns, calculate the e.m.f. induced in it if the current in A changes at the rate of 200 A/s.   *Ans* 10 V

Chapter 12

1. A parallel plate capacitor is arranged so that it can have air, paraffin wax of relative permeability 2.2, or glass of 8 as its dielectric. If the capacitor is charged to the same voltage in all three cases compare the quantities of electricity.   *Ans* 1:2.2:8

2. If the capacitor in example 1 is given the same charge when each dielectric is used, compare the potential differences between the plates.
                                                    Ans 1:0.454:0.125

3. A parallel plate capacitor has plates each of area 1 m², the separation being 5 mm. Calculate the capacitances (a) with air as dielectric, (b) with a medium of relative permittivity 3.
                                    Ans (a) 1770 pF; (b) 5310 pF

4. The capacitor in example 3 has a p.d. of 10 000 V applied to the plates; calculate in each case the quantity of electricity, and the electric flux density.
   Ans (a) 1.77×10^{-5} C; 2×10^6 V/m; (b) 5.31×10^{-5} C; 2×10^6 V/m

5. A parallel plate capacitor is made up of a number of separate plates each 1 m square, the separation being 1.3×10^{-3} m in air. If the total capacitance is to be not less than 0.4 mF and if the outside plates are of the same polarity, how many plates are required?   *Ans* 61

6. A capacitor of 30 mF is joined in series to one of 45 mF and a p.d. of 200 V applied Calculate (a) the charge on each capacitor, (b) the p.d. across each capacitor.   *Ans* (a) 3.6×10^{-3} C; (b) 120 V and 80 V

7. There are three capacitors of 1, 2 and 3 mF respectively. What are the total capacitances when they are connected (a) in series, (b) in parallel?
                                    Ans (a) 0.545 μF; (b) 6 μF

8. If a p.d. of 1000 V is applied to the capacitors in example 7 calculate the quantity of of electricity (a) when in parallel, (b) when in series, (c) also the p.d. across each capacitor in the series arrangement.
            Ans (a) 6×10^{-3} C; (b) 5.45×10^{-4} C; 545, 273 and 182 V

9. Two circular plates, each of diameter 20 cm, are facing one another 0.5 cm apart. If the relative permittivity of the dielectric between them is 2.5 calculate the capacitance.   *Ans* 139×10^{-12} F

10. A capacitor of 0.01 μF is charged to a p.d. of 200 V. Calculate the energy stored in it.   *Ans* 0.0002 J

11. A parallel plate capacitor has a dielectric 3 mm thick and the dielectric replaced by another of relative permittivity $3\frac{1}{3}$. Calculate the ratio of the second to the first capacitor.   *Ans* 1:3

12. A 10 μF capacitor is in series with a 10 MΩ resistor and a p.d. of 100 V applied. Plot the curve of charging current against time and from it determine the instantaneous values of the charging current at 0.05, 10 and 100 s after the application of the p.d.
                                    Ans 100, 60 and 14.0 μA

Chapter 13

1. The atomic weights of hydrogen, oxygen and silver are as 1, 16 and 108 respectively and of these only oxygen is divalent. Given that the electrochemical equivalent of silver is 0.001118 g/C calculate the values for hydrogen and oxygen.
   Ans 0.00001035 g/C for hydrogen 0.0000828 g/C for oxygen

2. If a current of 6 A deposits 8 g of a metal in 20 minutes, what is the electrochemical equivalent of the metal?   *Ans* 0.00111 g/C

3. It is required to deposit 0.5 g of metal of electrochemical equivalent 0.000304 g/C. If the p.d. at the terminals of the plating bath is 20 V and the resistance of the bath 150 Ω, how long will it take?
   Ans 205 min

4. Instead of using an ammeter to measure the current to a copper electro-plating cell, there is in series a resistance of 5 Ω and a voltmeter connected across this resistance. Current is passed for one hour, the reading of the voltmeter being 9.25 V. The mass of copper is 2.186 g. What value for the e.c.c. of copper does this data give?
   Ans 0.0003297 g/C

5. It is required to copper-plate a plaster cast. For this purpose the surface of the cast is brushed with very finely divided graphite so as to give a good conductivity surface. Given that the surface area is 100 cm^2, the thickness of plating required 1/20 mm, and the density of copper 8.8 g/cm^3, how many coulombs will be required? The e.c.c. of copper is 0.000329 g/C.   *Ans* 13 400 C

6. Given that 96 500 C liberate 1 g of hydrogen, how long will it take a current of 10 μA to deposit 1 g of (a) silver of atomic weight 108, (b) divalent copper of atomic weight 63.57?
   Ans (a) $24\frac{5}{6}$ hours; (b) $84\frac{1}{3}$ hours

7. It is required to copper-plate a metal sheet of 20×20 cm. If a current of 5 A is passed for one hour, calculate the thickness of the deposit, e.c.e. of copper 0.000329 g/C.   *Ans* 0.00089 cm

8. The current to a copper voltmeter is kept constant at 6.2 A according to an ammeter, for one hour. If 7.36 g of copper is deposited, what is the error of the ammeter?   *Ans* 0.03 A low

9. A copper voltameter and a wire of resistance 28 Ω immersed in 350 g of water are in series and a current passed through them for 18 minutes. If 0.86 g of copper is deposited in that time, what will be the rise in temperature of the water if the process is continued for half an hour?   *Ans* 20.2° C

10. For how long must a current of 20 A be passed through a solution of copper sulphate in order that 5 g of copper may be deposited in the cathode?   *Ans* 12.5 min

11. A battery of 50 secondary cells each of e.m.f. 1.8 V and resistance 0.05 Ω is being charged from a generator of terminal voltage 120. If

298   BASIC ELECTROTECHNOLOGY

the connecting leads have a resistance of 0.25 Ω but there is no other resistance in the circuit, what will the charging current be (a) at the commencement, (b) when the e.m.f. of each cell has risen to 2.2 V? What will be the p.d. at the terminals of each cell in the two cases?

Ans (a) 10.9 A, 2.35 V; (b) 2.38 V, 3.64 A

12. If, in the above problem, the generator p.d. is 200 V and it is desired to keep the charging current constant at 20 A by including a variable resistor in the circuit, what must be its resistance (a) at the commencement, (b) at the end of the charge? *Ans* 3 Ω; 1.75 Ω

13. A battery of 60 cells in series is to be charged from a 220 V supply. The e.m.f. per cell at a certain stage is 2 V. The resistance of each cell is 0.03 Ω and of the connecting leads 0.2Ω. If a controlling resistor of 20 Ω is used, calculate (a) the charging current at this stage, (b) the power wasted in the resistor, (c) the power actually required by the battery itself. *Ans* (a) 4.54 A; (b) 413 W; (c) 545 W

14. Two groups of cells A and B are to be charged in parallel from a 200 V supply, with a controlling resistor R connected in series with the parallel combination. In series with battery B there is also a controlling resistor r. Battery A has a total e.m.f. of 120 V and resistance 1.5 Ω, while battery B has an e.m.f. of 100 V and resistance 1 Ω. Calculate the resistances of R and r if A is to be charged at 6 A and B at 4 A. *Ans* R = 7.1 Ω; r = 6.25 Ω

15. A certain cell has an e.m.f. of 2 V and resistance 1 Ω. Another cell has an e.m.f. of 2.5 V and resistance 2 Ω. A certain resistance takes the same current whether connected to the first cell or to the second. What is its value? *Ans* 3 Ω

16. Each cell of a battery of 55 has an e.m.f. of 2.05 V and resistance 0.001 Ω. Each cell has 10 positive and 11 negative plates of active area 144 in^2 per side. If the current is 0.04 A per in^2 of positive active surface, what will be the p.d. at the terminals of a load supplied via a cable of total resistance 0.017 Ω. *Ans* 104.5 V

17. A lead-acid secondary cell has one positive and two negative plates, each plate having an active area of 3 × 4 in. If the plates are ½ in apart and the resistivity of the acid is 1.3 Ω cm, calculate the resistance of the cell. *Ans* 0.0105 Ω

Chapter 14

1. The positive half-wave of an alternating current can be plotted from the following data:

Time (s)	0	0.001	0.002	0.003	0.004	0.005	0.006	0.007
Current (A)	0	2.1	3.2	4.1	5.1	6.9	7.2	6.2

Time (s)	0.008	0.009	0.010
Current (A)	4.4	2.5	0

Plot the complete wave, the two halves being identical in shape. Determine the maximum value and the frequency.

Ans 7.4 A, 50 Hz

2. During one half-wave an alternating current has the following instantaneous values at equal intervals of time: 0, 5, 11.2, 17.5, 21.2, 23.2, 24.5, 25, 24.2, 23.5, 24.2, 25, 24.2, 21.7, 18, 13.5, 9, 4, 0. Plot the half-wave and calculate its r.m.s. and average values.

Ans 19.2 A and 17.4 A

3. What are the periodic times for frequencies of 15, 25, 50, 60 and 100 Hz?   *Ans* 0.0667, 0.04, 0.02, 0.0167 and 0.01 s

4. What are the r.m.s. and average values of a sinusoidal alternating voltage of maximum value 150 V?   *Ans* 106 and 95.5 V

5. A sinusoidal current of frequency 50 Hz has a maximum value of 10 A. If time is reckoned from when the current is zero and is becoming positive, calculate (a) the instantaneous current after 4 ms, (b) how long is required for the current to reach three-quarters of its maximum value for the first time?   *Ans* (a) 9.51 A, (b) 2.70 ms

6. A sinusoidal alternating current of frequency 50 Hz has an r.m.s. value of 20 A. Reckoning time from the instant the current is zero and is becoming positive, calculate the instantaneous current (a) after 1/400 s, (b) after 19/600 s.   *Ans* (a) 20 A, (b) −14.14 A

7. An alternating current of rectangular waveform and one of sinusoidal form have the same frequency and the same maximum value. Calculate the relative amounts of heat when flowing in circuits of the same resistance for the same time.   *Ans* 2:1

8. Two sinusoidal voltages of r.m.s. values 200 V and 300 V and of the same frequency, act together in the same circuit, the larger lagging the smaller by 60°. Calculate the resultant voltage and its phase with respect to the smaller component.   *Ans* 435 V lagging by 33° 36′

9. A cable supplies three separate motors which take the following currents: 20 A in phase with the voltage, 25 A leading by 20°, and 35 A lagging by 30°. Calculate the total current and its phase with respect to the voltage.   *Ans* 74.5 A lagging 7°

10. An inductive coil of negligible winding resistance takes 12.5 A when connected to a 200 V, 50 Hz supply. What is its inductance?

Ans 0.0509 H

11. A coil of 0.07 H and negligible winding resistance is in parallel with a pure resistance of 22 Ω. What is the total current if the supply is at 200 V, 50 Hz?   *Ans* 12.85 A lagging 45°

12. A capacitor of 50 μF is connected to a 110 V supply. What are the currents for frequencies of 25 and 10 000 Hz? *Ans* 0.864 A, 345.6 A

13. A resistor is required to take a current of 10 A at a p.d. of 35 V. If the supply is 100 V at 50 Hz what must be the inductance of an inductor connected in series with it so as to limit the current to the required value? What will be the inductor voltage drop?   *Ans* 0.03 H, 94 V

14. A p.d. of 100 V at 50 Hz is applied to a *RL* circuit with $R = 1.25 \, \Omega$ and $L = 0.07$ H. Calculate the current and its angle of lag.
             Ans 4.5 A, 84° 45′

15. When a certain inductive coil is connected to a 100 V d.c. supply it takes 10 A. When connected to a 100 V, 50 Hz supply it takes 5 A. What is the inductance of the coil?   *Ans* 0.055 H

16. A resistor of 30 Ω is in series with a capacitor of 79.6 μF. If supplied at 200 V, 50 Hz, calculate the current and its angle of lead.
             Ans 2 A, 53° 8′

17. An alternating p.d. of constant r.m.s. value 110 V is applied to a *RL* series circuit. When the frequency is 40 Hz, $I = 19.7$ A, and when it is 80 Hz, $I = 15.6$ A, calculate R and L.   *Ans* 5.0 Ω; 8.01 H

18. When a resistor of 200 Ω is in parallel with a certain capacitor and a p.d. of 250 V, 50 Hz applied the total current is 2 A. Calculate the capacitance.   *Ans* 19.9 μF

19. A p.d. of 550 V at 50 Hz is applied to a circuit consisting of a capacitor of 4 μF in parallel with a resistor of 625 Ω. Calculate the current and its angle of lead.   *Ans* 1.12 A, 38° 11′

20. At 500 V, 50 Hz, a certain coil takes a current of 20 A which lags 41° 25′. What will be the lag, or lead of the current if the coil is connected in parallel with a capacitor of 84.2 μF?   *Ans* 0°

21. If the supply is at 50 Hz, what capacitance connected in parallel with a coil of resistance 1 Ω and inductance 0.05 H will cause the total current to be in phase with the supply voltage?   *Ans* 199 μF

22. What current will flow in the *RL* circuit of problem 17 if the frequency is 120 Hz?   *Ans* 12.1 A

23. A *RLC* series circuit has $R = 100 \, \Omega$, $L = 2$ H, $C = 100$ μF. If the applied p.d. is 700 V at a periodicity $(2 \, \pi f)$ of 100. Calculate the current and its phase.   *Ans* 4.94 A lagging 45°

24. A certain indictive coil takes 15 A from a 230 V a.c. supply and the power is 1300 W. Calculate the angle of lag of the current behind the applied p.d.   *Ans* 67° 55′

25. A p.d. of 200 V at 50 Hz is applied to a *RL* series circuit with $R = 5 \, \Omega$ and $L = 0.01$ H. Calculate the power intake.   *Ans* 5720 W

26. A *RL* series circuit having $R = 10 \, \Omega$ and $L = 0.02$ H is connected to a 100 V, 50 Hz supply. Calculate the current, its phase and the power intake.   *Ans* 8.48 A lagging 32°; 720 W

27. A resistor of 1 Ω is in series with a variable inductor. If the supply is at 100 Hz, calculate the inductance necessary to reduce the current to (a) $\frac{1}{2}$, (b) $\frac{1}{4}$ of the current taken by the resistor alone. Calculate also the relative power intake for the two cases.

Ans 2.67 and 6.27 μH, $\frac{1}{4}$ and $\frac{1}{16}$

28. A *RL* series circuit has a power factor of 0.866 where $f = 50$ Hz. If $R = 5$ Ω what are the values of L and Z?   *Ans* 0.0092 H, 5.77 Ω

29. A *RC* series current has a power factor of 0.5 where $f = 50$ Hz. If $R = 5$ Ω, what are the values of C and Z?   *Ans* 368 μF; 10 Ω

30. A coil of $R = 2$ Ω and $L = 0.05$ H is in parallel with a capacitor of $C = 300$ μF. Calculate the power-factor of the whole when the frequency is 50 Hz.   *Ans* 0.247 leading

31. A *LC* series circuit is to have a natural frequency of oscillation of 830 kHz. If $C = 0.0002$ μF what must be the value of L?   *Ans* 184 μH

32. A *LC* series circuit is to have a natural frequency of 356 Hz. If $L = 0.02$ H what must be the value of C?   *Ans* 10 μF

33. A *RLC* series circuit has $R = 6.7$ Ω, $L = 0.54$ H and $C = 6$ μF. Calculate the frequency for resonance and the current, if the applied p.d. is 220 V at this frequency.   *Ans* 88.6 Hz; 32.7 A

34. A *RLC* series circuit has $L = 0.05$ H and $C = 20$ μF. Calculate the current and the volt-drops across inductor and capacitor if the supply voltage is 100 V at the resonant frequency of the circuit (a) if $R = 2$ Ω, (b) if $R = 20$ Ω.   *Ans* (a) 50 A, 2500 V, (b) 5 A, 250 V

Chapter 15

1. A three-phase motor has an output of 250 hp, its efficiency then being 92 per cent and power factor 0.89. If the supply voltage is at 2200 V calculate the line current.   *Ans* 60 A

2. A star-connected appliance takes 5 kW from a 3-phase, 440 V supply, its power factor being 0.8. Calculate the line current and the phase voltage.   *Ans* 8.2 A, 253.6 V

3. A 230 V, 3-phase, delta-connected motor has a power-factor of 0.88. When its putput is 20 hp its efficiency is 0.85. Calculate the phase current.   *Ans* 50.3 A

4. A balanced 3-phase load is supplied at a line voltage of 400 V. If the load is 8 kW at a power factor of 0.7 lagging calculate the load phase current (a) if it is star-connected, (b) if it is delta-connected.   *Ans* 16.5 A; 9.5 A

5. Two balanced 3-phase loads, A and B are supplied at 400 V. A is delta-connected and takes 5 kW at a power factor of 0.7 lagging, while B is star-connected and takes 6 kW at unity power factor. Calculate the total line current, the overall power factor and the phase currents in A and B.   *Ans* 17.5 A; 0.9 lagging; 6 A; 8.66 A

Chapter 16

1. A single-phase transformer steps down from 3000 to 600 V. Its output is 50 kW at a power factor of 0.8 lagging. Assuming 100 per cent efficiency calculate the approximate primary and secondary currents.

    Ans $I_1 = 20.8$ A; $I_2 = 104.2$ A

2. A 100 kV A single phase transformer steps down from 2000 to 400 V. There are 44 secondary turns, the gross core section is 453 cm² and the stacking factor is 0.85. If the frequency is 50 Hz, calculate max. *B*.

    Ans 1.09 T

3. A small transformer of 10 kV A output has an induced voltage of 5.1 V per turn. If the iron is worked at a maximum flux density of 1.1 T calculate the net iron section of the core.   *Ans* 51 cm²

4. The secondary winding of a transformer has 50 turns. The induced voltage on open circuit is 100 V at 50 Hz. The cross-section of the core (net iron section) is 129 cm². What is the maximum value of the flux density in the core?

    Ans 0.706 T

5. A 10 kV A transformer steps down from 2300 to 230 V. The primary and secondary resistances are 4.14 and 0.0472 Ω respectively. Calculate the total resistance referred to the secondary.   *Ans* 0.0886 Ω

6. The reactance of the above transformer is 4 times the resistance. Calculate the secondary terminal voltage when delivering full-load current (a) at unity power-factor, (b) at 0.8 lagging, (c) at 0.8 leading.

    Ans (a) 226.2 V, (b) 217.7 V, (c) 236.2 V

7. A 500 kV A transformer steps down from 2000 to 500 V. The secondary and primary resistances are 0.007 Ω and 0.17 Ω respectively. The core loss is equal to 15 kW. Calculate the efficiency when delivering full-load current (a) at unity power-factor, (b) at a power-factor of 0.8 lagging.   *Ans* 90 per cent, 92.2 per cent

8. If the above transformer is delivering one half of its rated current, what are the efficiencies at the same power-factors?

    Ans 92.7 per cent and 91.0 per cent

9. A 12.5 kV A single-phase transformer steps down from 440 V to 220 V. When on no load it takes a current of 2 A at a power-factor of 0.4 lagging. What is the core loss? The resistances of the winding are primary 0.35 Ω and secondary 0.08 Ω. Calculate the total resistances referred to the secondary and hence calculate the total loss at (a) full load current, (b) half-load.   *Ans* 352 W; 0.1675 Ω; 892 W; 487 W

10. A 50 kV A transformer has a core loss of 300 W and a total full-load copper loss of 450 W. Calculate its efficiency (a) at full load, unity power-factor, (b) at half load, 0.8 power-factor lagging.

    Ans 98.5 per cent, 98 per cent

11. A 10 kV A transformer steps down from 3300 to 240 V. The primary and secondary resistances are 6 Ω and 0.04 Ω respectively and the

reactances are 2.5 times the resistance in each case. Calculate the secondary terminal voltage when the transformer is delivering full-load current (a) at unity power-factor, (b) at 0.6 power-factor lagging.

Ans (a) 237 V; (b) 232.2 V

12. A 3-phase transformer steps down from 3300 V to 600 V. It is connected delta-star. If its output is 200 kV A, the power-factor 0.7 and the efficiency at this loading 97 per cent, calculate the line-currents and phase currents and the phase voltages on both primary and secondary sides.

Ans $I_1 = 36.15$ A, $I_{p.1} = 20.8$ A; $I_2 = I_{p.2} = 192$ A; $V_{p.2} = 347$ V

Chapter 17

1. A rectangular coil of area $\pi \times 10^{-2}$ m^2 has 1000 turns. It is rotated at a speed of 3000 rev/min in a uniform field of 0.5 T. What is the maximum value of e.m.f. induced in the coil? *Ans* 4930 V

2. A circular coil of diameter 1 cm and having 20 turns is rotated at 600 rev/min in a uniform magnetic field of 0.2 T. Calculate the average and maximum values of the induced e.m.f.

Ans 1.257×10^{-2} and 1.97×10^{-2} V

3. Calculate the frequency of the induced e.m.f. of a 12-pole alternator of speed 500 rev/min. *Ans* 50 Hz

4. In a 14-pole alternator how many cycles are undergone during three-quarters of a revolution? What must be the speed to give a frequency of 50 Hz? *Ans* $5\frac{1}{4}$; 428.6 rev/min

5. A 2200 V star-connected, 50 Hz alternator has 12 poles. The stator has 108 slots each with 5 conductors. Calculate the necessary flux per pole to generate 2200 V on no-load. Take the value of k as 1.07.

Ans 0.066 Wb

6. A 12 000 kV A, 50 Hz, 150 rev/min alternator has a star-connected, 3-phase winding housed in 360 slots. There are four conductors per slot. Calculate the generated e.m.f. when the flux per pole is 7.55×10^{-2} Wb. Take $k = 1.07$.

Ans phase voltage = 3800 V; terminal voltage = 6600 V

7. When a shunt generator is driven at its normal speed and the field current varied over a wide range, the following values of the induced e.m.f. were observed:

Field current

| 0 | 0.5 | 1.0 | 1.5 | 2.0 | 2.5 | 3.0 | 3.5 | 4.0 | 4.5 | 5.0 | 5.5 |

Induced volts

| 3.5 | 18.5 | 35.4 | 53.5 | 72.0 | 87.5 | 97.0 | 105.2 | 111.7 | 116.5 | 121.0 | 123.0 |

Plot the induced volts against field current. Draw the resistance lines for field current resistances of 31.8 Ω and 24 Ω and hence, from their intersections with the magnetisation characteristic determine the generator voltages when the field circuit resistance has these values.

Ans 100 V and 120 V

8. A d.c. machine has an armature resistance of 0.6 Ω. When the armature current is 30 A and the terminal voltage is 500 V what will be the induced e.m.f. (a) when acting as a generator, (b) as a motor.

Ans (a) 518 V; (b) 482 V

9. A 4-pole generator with wave wound armature has a flux per pole of 6.4×10^{-2} Wb. There are 450 armature conductors. What is the induced e.m.f. when the speed is 500 rev/min? *Ans* 480 V

10. A 550 V generator rotates at 330 rev/min. The armature is lap wound and has 960 conductors. What is the flux per pole? *Ans* 10.4×10^{-2}Wb

11. A compound generator giving 600 V at no load supplied a load via a cable whose cores are each of 0.153 Ω resistance. When loaded to 250 A what must be the generator terminal voltage in order that the voltage at the load may remain at 600 V. *Ans* 676.5 V

12. A 4-pole generator with wave-wound armature has 51 slots each having 48 conductors. The flux per pole is 7.5×10^{-3} Wb. At what speed must the armature rotate to give an induced e.m.f. of 440 V?

Ans 7.9 rev/min

13. The magnetisation characteristic of a d.c. shunt generator when run at the normal speed is:

Field current	0.5	1.0	1.5	2.0	2.5	3.0	3.5
Induced e.m.f.	60	120	138	145	149	151	152

What must be the resistance of the field current in order that the induced e.m.f. may be 150 V? *Ans* 53 Ω

14. A d.c. armature winding has 192 turns each of length 40 in the wire having a cross-section of 0.02 in². The machine has four poles and the armature is wave wound. Calculate the armature resistance, given that $\rho = 0.76 \times 10^{-6}$ Ω in. *Ans* 0.0729 Ω

15. If the winding in problem 14 is worked at a current density of 2500 A/in², calculate (a) the volt drop in the windings (b) the I^2R loss.

Ans (a) 7.3 V; 730 W

16. A 220 V d.c. motor has an armature resistance of 0.75 Ω. What is the back e.m.f. when loaded to an armature current of 20 A?

Ans 205 V

17. A 220 V shunt motor has an armature resistance of 0.5 Ω and a shunt field resistance of 100 Ω. If the load is such that the back e.m.f. is 210 V, calculate the armature and line currents. *Ans* 20 A; 22.2 A

18. A 500 V shunt motor has an armature resistance of 0.6 Ω and a shunt resistance of 750 Ω. On no-load the line current is 2.67 A and the speed 1200 rev/min. What is its speed when loaded to a line current of 40.67 A? *Ans* 1148 rev/min

19. A 200 V, 4-pole shunt motor with lap wound armature has 840 armature conductors, the resistance being 0.25 Ω. The flux per pole is 2.5×10^{-2} Wb. Calculate the speed when loaded to an armature current of 100 A. *Ans* 500 rev/min

20. With the armature current of the above motor remaining at 100 A what additional resistance in series with the armature will (a) reduce the speed to 250 rev/min, (b) bring the motor to standstill?
Ans (a) 0.875 Ω, (b) 1.75 Ω

21. A 500 V d.c. series motor has a total resistance of 0.9 Ω. When the load is such that the current is 45 A the speed is 12.5 rev/s. Assuming that the magnetic circuit is not saturated and that the flux per pole is therefore approximately proportional to the current, what will be the speed when loaded to 60 A? *Ans* 9.1 rev/s

22. A 1000 hp 50 Hz induction motor has 12 poles. At full load its slip is 1.5 per cent. What is its speed? *Ans* 492.5 rev/min

23. What is the frequency of the rotor induced e.m.f. of the above motor at full load? *Ans* 0.75 Hz

24. A 6-pole, 3-phase, 400 V, 50 Hz induction motor has a full-load output of 8 hp, its efficiency being 84 per cent and power-factor 0.75 lagging. Calculate the current taken from the supply and the speed. *Ans* 13.8 A; 960 rev/min

25. A triple ram pump is driven by a 16-pole, 50 Hz induction motor. The motor runs with a slip of $2\frac{1}{2}$ per cent, and the ratio of the gearing between motor and pump is 7.5 to 1. Calculate the speed of the pump.
Ans 48.8 rev/min

Chapter 19

1. A moving iron ammeter has a coil requiring 400 ampere turns to give a full-scale deflection. How many turns will be required if the instrument is scaled to (a) 100 A, (b) 0.5 A? *Ans* (a) 4; (b) 800

2. How many turns will the coil of an instrument similar to the above require if it is to be used as a voltmeter scaled to 100 V and taking a current of 0.04 A? What series resistance will be required in a multiplier so that the instrument may measure voltages up to 300 V?
Ans 10 000 turns; 5000 Ω

3. A voltmeter having a total resistance of 500 Ω is scaled to 2 V. What series resistance is required to increase the range to 100 V.
Ans 24 500 Ω

4. The moving coil of a certain instrument gives a full-scale deflection with a p.d. of 75 mV, the current then being 25 mA. What values of shunt are required for ranges of (a) 0–5 A, (b) 0–50 A.
Ans (a) 0.01508 Ω; (b) 0.001501 Ω

5. The movement of a permanent-magnet moving-coil instrument has a resistance of 12 Ω and gives full deflection when the current is 5 mA. Calculate (a) the shunt resistance for use as a 0–5 A ammeter, (b) the multiplier resistance for use as a 0–100 V voltmeter.

Ans (a) 0.012 Ω; (b) 19.988 Ω

6. A galvanometer has a resistance of 100 Ω. What shunt resistance is necessary if the galvanometer is to carry only $\frac{1}{10}$ of the total current? If the total resistance is to be unaltered, what series resistance will be necessary when the shunt is used?   *Ans* 11.11 Ω; 90 Ω

7. A milliammeter has a resistance of 5 Ω and gives a full-scale deflection with a current of 20 mA. What shunt resistance will be required to increase the range to 5 A. If the instrument has a temperature co-efficient of resistance of 0.001 Ω per Ω per C°, while that of the shunt is negligibly small, what percentage of error will be caused by a temperature rise of 10 C° when the 5 A shunt is in use?

Ans 0.02008 Ω; 1 per cent low

8. The resistance of a certain appliance is being measured by the ammeter and voltmeter method. The voltmeter resistance is 300 Ω and the ammeter resistance 0.002 Ω. The ammeter reads 10 A and the voltmeter 1 V when it is connected to the outside ammeter terminal.

    If the voltmeter is now connected to the inside ammeter terminal, that is, directly across the resistor being measured, what will be the percentage difference in the two calculated values of the resistance?

Ans 1.2 per cent higher

9. A metre bridge has for its slide wire two 50 cm lengths of different wires in series. If the resistance per cm of the second half is twice that of the first, where will the balance be obtained if the resistance in the first gap is 30 Ω and that in the second gap 2 Ω.

Ans 20 cm from the junction

10. In a Wheatstone bridge network ABCD, the cell is connected between A and C and the galvanometer between B and D. If $R_{AB} = 10$ Ω, $R_{BC} = 30$ Ω and $R_{AB} = 57.5$ Ω, what will be the resistance of DC when a balance is obtained? If R_{DC} is slightly less in value and the positive terminal is connected to A, in what direction will the current flow in the galvanometer?   *Ans* 17.25 Ω; B to D

11. The resistances in the arms of a Wheatstone bridge ABCD are $R_{AB} = 10\Omega$, $R_{BC} = 30$ Ω, $R_{AD} = 50$ Ω. The cell is connected to AC. If the e.m.f. of the cell is 1.5 V and its resistance 8 Ω, what current will the cell deliver when the bridge is balanced?   *Ans* 0.036 A

12. A Wheatstone bridge has ratio arms P and Q. The standard S is adjacent to P and the unknown X adjacent to Q. With $P/Q = 100$, the balance lies between $S = 9463$ and $S = 9464$. With the former value the galvanometer deflection is 36 and with the latter value it is 58 in the opposite direction. What is the value of X?

Ans 96.634 Ω

INDEX